THE NEW ZEALANDERS

To Jane, Danny & Tommy,

Many thanks for your very generous hospitality and for the kindness extended to a lone Kiwi.

Hope this gives you some idea of what we're all about and please remember —:

You'll always be welcome here.

Haere ra

Moana Herewini
(Hitchhiker '75)

Endpaper:
Tauranga Bay, Northland.

First published 1975
by Golden Press Pty Ltd
16 Copsey Place
Avondale, Auckland
and
35 Osborne Street
Christchurch
Printed in Hong Kong
ISBN 0 85558 332 0
© Robin Smith and Warren Jacobs

THE NEW ZEALANDERS

Robin Smith
Warren Jacobs
Graham Billing

Golden Press
Auckland · Christchurch · Sydney

FOREWORD

In its small compass New Zealand has a span of grandeur, tranquility and contrast rarely found anywhere else in the world. From the sub-tropical north to the permanently snow-capped Southern Alps the land has presented a challenge for generations of New Zealanders. The result of this challenge is a unique combination of land and people—a combination which our people have dedicated themselves to conserving for coming generations.

The land has given us the basis for our wealth; our people the labour and ingenuity to develop it. The country's diversity and our regard for this have moulded the character of New Zealanders. From this very diversity we are welding a unity among our peoples—Caucasian, Polynesian and Asian.

In the 1970s the goal of unity has become the spur to the development of a new confidence and vitality in our society. There is a mood of renewal and change in the cities and countryside. Our people are seizing the opportunity to participate, to make things happen through their own efforts. That in itself is a renewal of the New Zealand spirit.

Social and economic justice for everyone has been basic to our way of life since New Zealand was a fledgling society. In seeking that justice, New Zealanders have brought about a mix of individual and government action which is found nowhere else. It helps the New Zealander to help himself, to develop his own talents, roll up his sleeves and tackle the problems which confront him as a person or in the wider society.

The New Zealanders helps to portray the way of life which comes from mixing a diverse land and diverse peoples. It helps to portray us as today's *tangata whenua*—the people of this land.

W. E. Rowling
Prime Minister

CONTENTS

1 THE PEOPLE OF THE LAND　7

2 THE LANDBUILDERS　66

3 THE MIXTURE　90

4 THE TOWNS AND CITIES　110

5 THE ECONOMY　130

6 THE SCIENCES　154

7 THE ARTS　162

8 SPORTS　175

THE PEOPLE OF THE LAND

Burnt by the south wind a bewildered lemon tree grows in the back yard of Bob Buckley's farmhouse, and a single lemon glows in the winter darkness. Soft light from the kitchen window reveals the tree almost bare of leaves. The soil above its roots is frosted white but somehow, on a coast unkind to lemon trees, it will survive. The farmhouse sets its bulk sternly against the southerly winds that bring upon them that breath of the Southern Ocean which can seem to freeze an essence of the great whales' exhalations into the soil of kitchen gardens, the wool of sheep, meat hanging in the safe which swings from the sacred *karaka* tree, the very butter in the kitchen refrigerator. The little lemon has a red tinge on the face it shows to the sun, like a chilblain.

The house is on a small level place dug in the side of a hill, and the hill is the seaward slope of an ancient beach raised long ago above the sea by an earthquake. On this old beach, now millennia old, sheep fat with lamb and their own meat graze pastures rich with nutrients an aeroplane pours on them every spring. A water pump set in the stream waters of a gully beside the house beats like the still earth's heart. Sheep snicker in the scrub-dark faces of the gully, opossums scream. The land is spreading out in darkness to embrace the sea, an act in which it is always consumed.

The surf on the stony beach of Palliser Bay is for once quiet, hardly more than a wash up and down the stones. There is no sand here. The sea is too violent to let it rest. Here and there ridges of black sedimentary rocks jut into the sea and the night is quiet enough to hear fur seals breathing as they rest from a deep dive. They will not rest long. Some perturbation of the tide or current down the long chasm of Cook Strait which divides the main islands of New Zealand, North and South, will warn them of an inshore danger and they will be gone to safer rookeries.

Buckley is in his airing cupboard, the space around the electric hot water cylinder where his heavy woollen working underclothes and shirts and socks, wet from yesterday's cattle muster in the bush and wading streams in search of early lambs born to lost mothers, have dried out during his brief sleep. The kitchen is still warm from last night's open hearth fires. The first dog barks down at the creek and one or two others, doleful in their line of pens, whine morosely. Buckley puts the coffee percolator on the stove along with the porridge pot in which oatmeal has soaked overnight, a plain

On this land the giant moa *grazed. Men hunted him, dressed themselves in his feathers, made tools from his bones, drew him on rock in their first art. And when that bird died for the last time they found other birds to live on, and then four-footed animals like the pigs of the first white men coming, and more food like maize and potatoes, and then many white men came and the men became new, and the land changed, and they had the sheep for food and clothing. And then they cleared the land to make more room for sheep, and for all the people to grow in, and then they discovered how to freeze the meat and send it in great ships to other men 10 000 miles away, and they found how to take the wool and make their clothing in great quantity, for many men, all around the world, and then they said, when the sheep came down in the morning to have the wool taken, that they had found a great peace.*

Above:
'From an aeroplane at 610 metres you can see the sea on either side.' The Southern Alps, looking towards the Tasman Sea.

Opposite:
Called a living fossil, the friendly tuatara lizard.

Above right:
In the mists of evening, a classic mountain seems like the very force that keeps the Moon in orbit. Mount Egmont.

Centre right:
Mae Rudolph, waitress from Whangarei.

Opposite right:
A great river begins—the headwaters in the bush.

Far left:
Pumice dust makes a pale Maori. Bulldozer driver, volcanic plateau forest.

Opposite:
Remembrance is bitter. The Cenotaph Christchurch.
On this spot a Mayor of Christchurch cast out wreaths for anonymous Vietnam war dead, laid by anti-war protesters in an Anzac Day dawn.

Below:
Round rocks, boiled in the earth's lava. Stepping stones for children at Moeraki, North Otago.

breakfast but good for four hours' hard work. He thinks of when somebody might come for dinner, enough people to make it worth while thawing a haunch of venison and some of last autumn's quail, or going to catch a bucketful of *kura*, the freshwater crayfish, in the creek. He can hear a morepork calling in the huge pine tree which leans over the water tank up the hill, the melancholy *ruru* of the dish-flat face and huge *paua*-shell eyes as the Maoris used to carve him.

In the freezing world of Palliser Bay on a still August night he knows that the wild pigs will have come out of the scrub gullies to plough the fields with their snouts seeking tender plant roots and soft-shelled insects and the lonely new-born lamb to be devoured. The red deer will be out on the pastures grazing among the sheep because there was no storm yesterday to leave a windfall of fresh leaves and branches on the forest floor. He peers south-west out of the kitchen window, down across the Strait to the highlands of the Kaikoura Range in the South Island and the 2 930 metre peak Tapuaenuku, the weathervane mountain across the sea for the people who live on the North Island's south coast. He sees too, the weathered blue gum tree at the foot of the house hill, where the creek runs by and the old straw-and-mudbrick chimney of the first house in Palliser Bay has not quite crumbled away. The hearthstone is still there and Buckley is a chimney builder who has rebuilt the hearths in his house and in the old homestead house down by the sea.

Tapuaenuku vanishes while he looks. For a moment he thought he could make out the snow cirques of its final peaks, almost bronze-coloured under the moon and the smoky fumes of Wellington City drifting southwards from Port Nicholson and the Hutt Valley sixty kilometres to his west. There is southerly weather coming just as it was forecast and he must go to sea to lift his lobster pots. He looks at the single small lemon on the lemon tree and admires its certitude. The tree was there when he came and will be there when he dies or goes away because somehow the soil is kind enough to keep it, and somehow the wind will never quite freeze the sap from its branches, and the salt on the wind will never quite rime the small leaves to suffocation. The porridge is cooked.

The Landrover is in the big vehicle shed, across the garden from the lemon tree, with the outboard motor, fuel can, life jacket, anchor and line and box of knives and gear. As he drives down the track he sees the young cattle fattening on the flat beach paddocks moving towards the fence line where in the daylit morning he will fill their fodder boxes with hay. Now, with all the vigour of his lean body he must hitch the boat trailer to the Landrover and drive across the road to the launching site on the beach. The road goes nowhere but to the bleak Cape Palliser with its towering white lighthouse and the seal rookeries of its black rock ramparts besieged by an arrogant sea twenty kilometres further round the bay. The road serves homesteads of six other sheep stations and two small lobster fishing settlements huddling on the windswept beach, little houses of peeling paint-work with wooden shutters to fix on their windows against the gales.

Above:
Looking south towards the Cavalli Islands from Whangaroa, Northland, at 1830 metres.

Left:
The 'Mayor' of Martins Bay—and the only permanent inhabitant.

Opposite:
Cape Kidnappers gannet colony, Hawkes Bay.

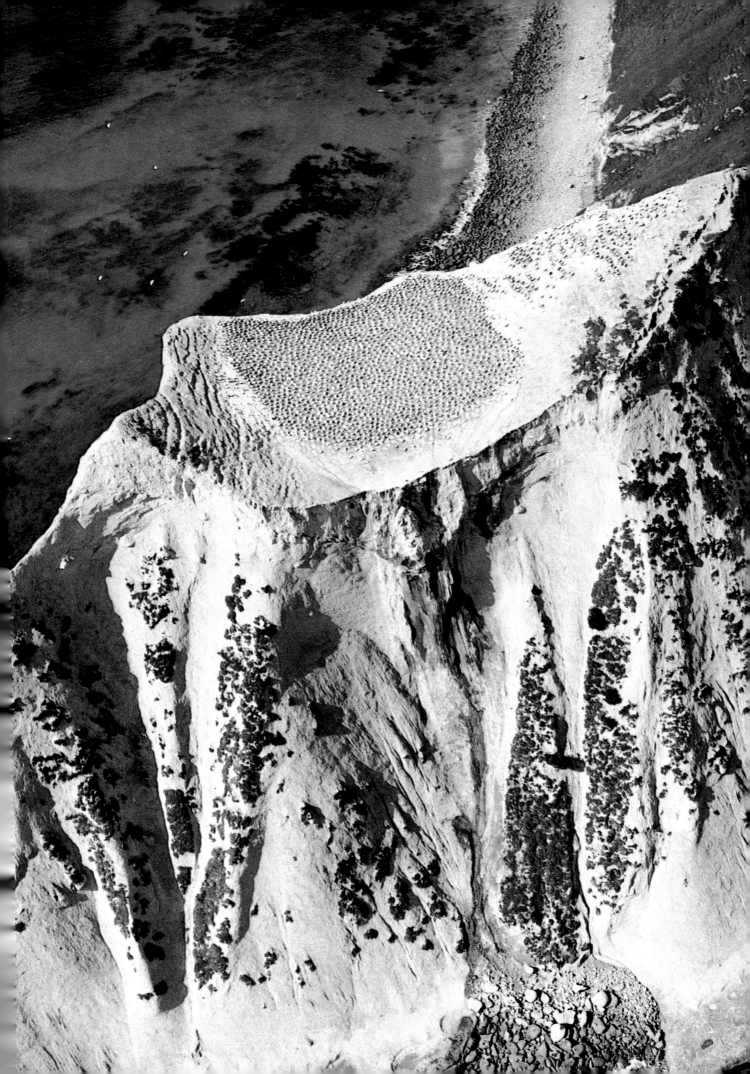

Opposite:
Shepherds and dogs, MacKenzie Country.
With just one marvellous dog, an illiterate Scots shepherd who spoke only Gaelic, James McKenzie, stole 1000 sheep from a South Canterbury run in 1855 and drove them through mountain country to Southland in one of the great New Zealand journeys of exploration.

Below:
Mount Ngauruhoe, one of three sisters who guard the wasteland of the central North Island.

At the Whatarangi Station gate a grader has recently been reforming the road, cutting into the banks with its blade and revealing stone adzes, knife blades and spear heads of obsidian from the Bay of Plenty volcanoes five hundred kilometres to the north of the island, middens of fossil sea shells from the now buried mudflats of Lake Wairarapa. For here some of the first people of New Zealand lived, the *tangata whenua* who discovered New Zealand, on their voyages from the northern islands of the South Pacific—the *tangata whenua* 'people of the land' as the Maoris of the great fourteenth century migrations called them, the ancient inhabitants who came to Palliser Bay as long as 2 000 years ago.

Above left:
Early morning, Crooked Reach, Stewart Island.

Opposite:
Rainbow over the woolshed at Lilybank Station in the Mackenzie Country.

Above:
Kotuku, the rare and sacred white bird whose name means 'Once in a Lifetime'.

Right:
Rangitoto Island at daybreak, silhouetted behind a delicate screen of toi-toi.

Opposite:
Pohutukawa, Piha.
Trees a thousand years old in whose shade the legendary lovers of the first Maori migrations intertwined.
Below:
Glow worm threads, Waitomo Caves.

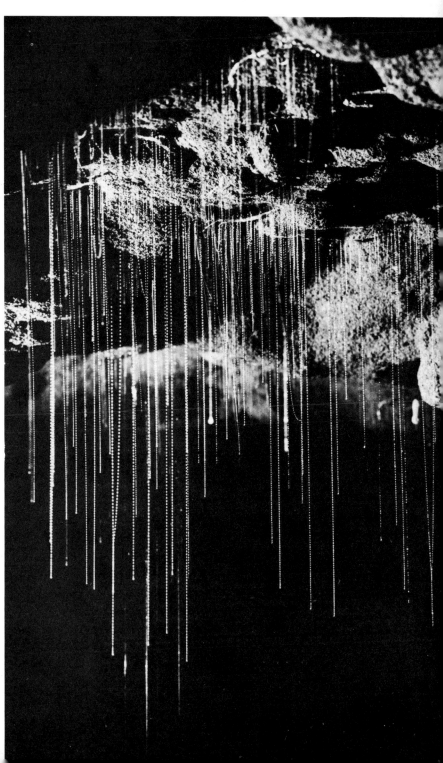

Opposite:
Bethell's Beach, Auckland's West Coast.

Opposite below:
A bach at the beach, Tongaporutu.
The weekend way of life.

Right:
Fishing for the sun? The salmon, traditional beast of wisdom, is caught at the Rakaia River mouth.

Above:
Mutton-birder with catch—sooty shearwater chicks, the traditional preserve of the Ngai Tahu tribe.
A national delicacy lies on the bush floor ready for processing—packed in eighteen-litre tins, preserved in salt and their own oil, they come from near sub-Antarctic islands.

They lived in pits dug in the ground. They built elaborate stone walls to shelter their *kumara* crops, the sweet potato they brought in their canoes from the warm islands of central Polynesia. The place was too cold for their seed crops to survive the winter so they built double stone walls, laid the seed *kumara* in the gap between, covered them with earth and kept them warm until spring by lighting fires on the surface. The stone buildings are still as mysterious as the Easter Island statues at the eastern point of the Polynesian triangle.

There were stone wall fortifications at Palliser Bay; stoneworks for a kind of open-air temple with an altar of stone on the bank of the Washpool Stream which marks one Whatarangi boundary; piles of stone marking ceremonial graves containing the bodies of people ritually killed with spear thrusts; hearthstones with green jade treasure ornaments, *tiki* and *mere* carved from *pounamu*, hidden under them; the centre posts of later houses, one found with a human skeleton still clinging to the base, teeth embedded in the wood with the force of the agony of ritual burial to bless the dwelling. There was the body of a great chief buried with his treasure and alongside him a child with a shell cloak still intact about him—more than a thousand years old and the shells of the cloak from a species which became extinct before the birth of Christ.

For Buckley, as he launches his aluminium boat in the grumbling wash of the beach, these things no longer seem strange. The Bay is not haunted. It simply has the certainty of 2 000 years of human history. People have used its land and sea in the way he does for two millennia. There is a certain violence in his confrontation with his world that is in an unbroken tradition. To protect his lobster pots from night time poachers he runs split razor blades into the lay of the last fathom of rope above his pots. They will lacerate the hands of any unsuspecting thief. The lobster tails he packs and freezes for export to America are part of the intricate chain of his livelihood and consecrated to the purpose of paying the mortgage on his 1 200 hectares of pastureland, scrub grazing, dense bush, steep gully and stream.

When today's catch is packed he'll feed hay to the black beef cattle, snatch breakfast eggs and coffee, and go to collect the bodies of the opossums who in the night have eaten poison bait from the line he runs along the homestead creek. He'll skin them and start the skins on their long journey to the fur markets of Britain and Europe. In the afternoon he will plan the muster of his 2 500 ewes-in-lamb for inoculation against the diseases of late pregnancy which threaten sheep and lamb alike. In the early evening he will drive up the rough bulldozed tracks to the bush and the hills behind the homestead and shoot the deer he knows will be grazing certain clearings at the edge of the forest trees. The venison will go to his freezer on its way to the sumptuous restaurants of West German cities. The antlers of the stags will go to Hong Kong to be made into Chinese aphrodisiacs along with other by-products producing strange tonics and medicines. The skins will go to a New Zealand tannery where they will be made into deer skin suede for the creation of men's and women's fashion garments. The money will pay the mortgage on the farm.

Opposite:
The world's most romantic place name.
In the Maori language the spoken name sounds as fluid and flute-like as the poem of its English meaning.
Below left:
A trophy rack. Red deer antlers on a West Coaster's garage.
Below right:
Stock sale 'yarder', Taranaki.

If the winter land is bitter here it is still full of riches to be harvested. When a tourist cruise ship berths at Wellington some of the men will come to Whatarangi to bag a red deer stag or a boar. They will sleep at the homestead after a gourmet meal which Buckley himself has prepared from his resources of fish and game. They will sleep deeply after a day in the solemn beach forests and the tangled tussock hills.

In summer, the little lemon tree will feel its struggle for survival justified and like the seed *kumara* kept alive under those ancient winter fires, will set down root and put on leaf. Like the European's lemon, the *karaka* tree which grows in many ancient groves in the steep gullies of the Palliser coast, was said in Maori tradition to have been brought to New Zealand by migrants. The chief Turi in the fourteenth century migration canoe *Aotea* brought *karaka* seed. Science doubts that this is true but the *karaka* groves did hold a mystic significance for the Maori and he carved dendroglyphs on their ancient boles, carvings of magic faces and totem animals. To the Maori, the *karaka* was a fruit tree bearing drupes from which the flesh could be eaten. The kernals, poisonous when fresh, yielded a nut-meal for winter storage after they had been baked and washed. For the white migrant of the last hundred years the lemon tree, with its origins in the Mediterranean cradle of European culture, has a significance just as ancient.

While Buckley, in his passion to preserve a way of life, struggles with history and the cold sea, New Zealanders up and down the 1600 kilometre arc of islands which makes up their country are looking for the same small signs of spring. In the world in microcosm that New Zealand is, they are abundant; some like the thawing and flowering of a tundra desert; some like the humid awakening of a rain forest from a monsoon season; some like the spring of prairie lands under the plough; some like the quickening of flowers under the slushy snow of an English county down land. On the Southland plains the South Island farmers are shearing their sheep hoping that some final blast of winter will not harass them before they drop their lambs. In the Canterbury high country, the tussock grassland of the foothills of the spectacular alpine ridge which plunges down the centre of the South Island, the run-holders are cursing spring snowfalls which have trapped their ewes in the drifts of shady mountain faces where they may die if there is no swift thaw.

In Central Otago, fruit growers who own the country's richest apricot and peach orchards are filling smoke pots with oil against the possibility of late frosts when the blossom comes. In Nelson Province tobacco and hop growers are praying for an early summer without thunderstorms, gales, and torrential rain. On the North Island's south-west coastal plains farmers watch the sky for an even rain to bring on the pasture growth that fattens lambs early and brings calving cows into fat milk. Their cousins over the central dividing ranges in Hawkes Bay and Poverty Bay are hoping that the hot nor'-westers of early summer will not blow too harsh and burn the pasture grass before it flourishes again in the new season.

Previous page:
A sunburst, and an opening. Milford Sound and Mitre Peak, the land as it was seen by the seal catchers in 1792.

Opposite:
Owen Moriarty, a man of the land.
Taihape-born, he has spent a lifetime on the land, and now works on a Waikato thorough-bred stud farm.

Far right:
Theatre, first night. Fur coat, jewels, for a production at the Town Hall.

Below:
An intent audience at a Sunday afternoon outdoor rock concert

Far left above:
Goldminer's house amidst snow and hoar frost, Central Otago.

Far left below:
The kiwi, a shy night bird who uses his long bill to catch worms and insects.

Far left below:
The kiwi a shy night bird who uses his long bill to catch worms and insects.

Centre left:
Pohutu Geyser, Whakarewarewa, Rotorua.

Opposite:
A rainbow breaks through the gloom of a mountain storm.

Below:
In the rain forests of Fiordland trees grow straight. They take energy from up to 8900 millimetres of rain a year.

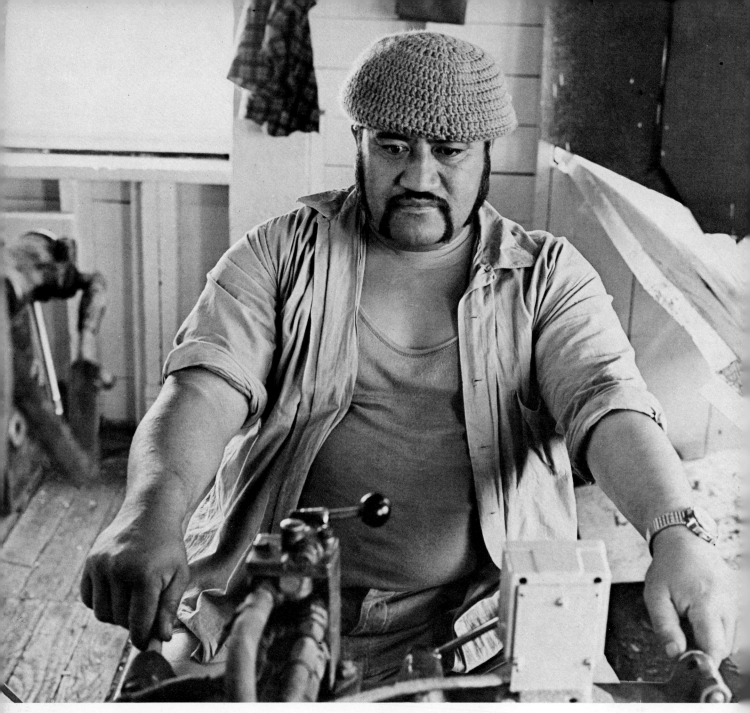

In the Bay of Plenty, orange growers pray for no spring gales which will wrench the last laden fruit from their trees. In the Waikato, fat lamb farmers hope for a spring cool but wet enough to grow the grass, not warm enough to multiply the fungal spores of deadly facial eczema dormant in their paddock soils, and for a good drop of heifer calves to replenish their herds rather than bull calves which must be slaughtered for veal. In North Auckland and the Coromandel the people of the land feel the seasons' shift gently. The climate does not become less cold, rather it becomes more pleasant.

And still, in the grim city of Invercargill, dawn comes late and factory workers' families huddle in their beds hoping that last night's hearth fire has retained its overnight coals which can be stirred into heat. In Wellington a warm, wet nor'-west wind may be creeping over the hills that form the amphitheatre of Port Nicholson, thrashing pollen from the suburban pine trees and winter flowering gorse. In Auckland the families of business and professional men begin to talk excitedly about the coming yachting season on the cruising waters of Hauraki Gulf, one of the world's finest sailing grounds. In the bleak South Auckland suburbs of worker housing like Otara and the Polynesian central city areas like Ponsonby, Cook Islanders, Niueans, Samoans, Tongans, and Tokaalauans think again of the tropical warmth of the island homes they have left.

At Tiwai Point near Invercargill the giant Comalco aluminium smelter continues without rest to devour the electricity generated by the South Island glacial lakes and fast blue rivers. At Glenbrook, south of Auckland, the New Zealand Steel Mills' furnaces devour ironsand from the Tasman Sea coast beaches, at Kawerau in the eastern Bay of Plenty the Tasman Pulp and Paper factories devour trees from the huge man-made forests that serve them.

New Zealand's island arc is long enough to encompass many climates though they are not strictly related to latitude. They are particular—Central Otago in the South Island has a desert climate perhaps like inland Algeria, small areas of 305–381 mm rainfall together with land that burns into bare earth, scab weed, flannel weed, foxglove and dry tussock grass heads in summer and may spend several weeks of winter under snow, at least months under frosts which keep the subsoil almost permanently frozen. In Southland the plains and downlands can spend weeks in rain, fog, snow storms and the blast of a bitter Antarctic wind.

Marlborough Province in the north of the South Island may meanwhile endure a five-year drought in which distraught farmers subscribe to funds to pay for seeding clouds to make rain and to

Opposite:
Sawing timber, New Zealand Forest Products, Kinleith.
'Maoris do have a role to play in New Zealand society and I have been through factories and noticed they make particularly competent and happy machine operators.'—The Hon. J. R. Marshall, former Prime Minister.

Below left:
Flowergirl, Samoan wedding, Auckland.
Pink for a little girl—but no plastic flowers for a migrant people who remember their earth and sea.

Below right:
Westland whitebaiter.

Above:
Mixed farming land, Canterbury Plains.

Left:
An icy landscape at Lake Lyndon has an other-world appearance.

Opposite above:
An evening storm over The Remarkables, at Queenstown.
First explored in 1859, The Remarkables were named for spectacular rock formations like the Devil's Staircase. Double Cone rises to 2344 metres.

Opposite:
An arm of the sea, Stewart Island.

Above:
Christ's College schoolboys, Christchurch.
From a by-gone age, in uniforms transplanted from Victorian England, boys mix scholarship with style.

Above opposite:
Red balloons, Christmas holiday camping at Mount Maunganui.

Opposite:
Commuters leaving work, Lambton Quay, Wellington.
Out of the 'street canyon' country of Wellington's Government Centre. At five o'clock, public servants head for commuter trains and dormitory suburbs.

import rain-making experts from Australia to advise them. Wellington Province just across Cook Strait from Marlborough, the province with the country's highest sunshine hours a year, suffers from salt-laden north-westerly gales for long seasons. They bring the capital city its nickname of 'Windy' Wellington. The southern hills of Wellington Province are bare and seared by the north-west and south-west winds; scrub trees grow bent, their foliage clipped by the wind into hunched shapes.

In the Island's far north there is warmth and peace broken only by the eccentric path of cyclones from the Coral Sea which wander far enough south to wash the Northland kauri forests, farms and orchards with tropical rain at their spinning edge. Within the microcosm the pieces of climate give the illusion that whole new states and lands exist here.

This particularity of climate and the pattern of animal and vegetable life it imposes, are reflected in the character of New Zealanders in subtle ways. Maoris and the modern Polynesian migrants who follow the sea roads explored by their ancestors for the past 2 000 years, settle in the warm north and the city of Auckland, capital of the South Pacific nations. Mixing with Europeans they create a different culture from that of the far south where in cities like Invercargill and Dunedin the Scottish, North-of-England and Irish migrant stock have developed a cold-climate culture and attitudes to social affairs which are bewildering to the casual tourist. At one end of the country there is a zesty air of freedom in the sun; at the other, a dour air in which the Protestant ethic proclaims that man is spiritually enhanced by his struggle with the elements. The land itself has helped to shape these differences.

Above:

In the depths of this water a Whakarewarewa child can see a copper coin. A game played by tourists and children alike.

Left:

Boiling mud, Tikitere.
A place name meaning 'my youngest daughter has floated away'.

Right:

Mini-volcano, Whakarewarewa mudpool, Rotorua.

Far right:

The earth was rent, flame and stone and ash came out, and mud, which killed 153 people in the Tarawera eruption of 1886.

Opposite:
Summer Sunday in the sun, Rupaki Bay, Banks Peninsula.
Right:
Roadside vegetable stall, Hawkes Bay.
Below:
It's a great game. Country sports day.

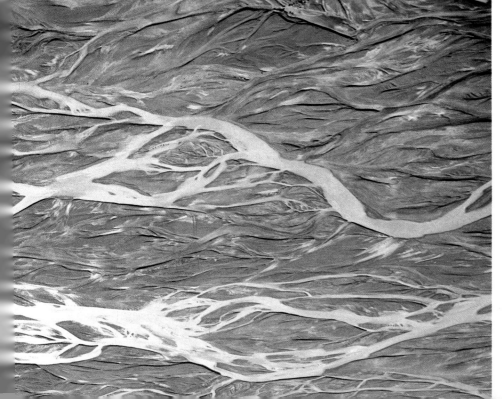

Far left:
Morning mist, Hagley Park, Christchurch.

Above:
Storm rising, Lake Hayes, Central Otago.

Opposite:
Riverbed pattern, Tasman River, Westland. The Tasman River at this point is more than two kilometres wide.

New Zealand is built from a great chunk of ancient rock of continental thickness. It lies along the circum-Pacific belt of volcanic activity. Australians on their stable continent 2 100 kilometres west across the Tasman Sea, refer to New Zealand as 'the Shaky Isles' and the jibe is warranted. The geological forces which have folded, buckled, twisted and sheared the sea floor to raise New Zealand above the waves, are still at work. Earthquakes, since European colonization began in the 1840s have drastically re-shaped some local landscapes like Murchison on the South Island's west coast or Napier in the North Island's Hawkes Bay. The great Wellington earthquake of 1855 raised shoreline platforms 1.5 metres out of the sea. Continuing activity, particularly on a broken line which runs from Fiordland in the south-west of the South Island to the Bay of Plenty and its offshore volcanoes in the North, a line of faults and fractures in the depths of the land, allows the spectacular volcanoes of the North Island's Ruapehu group to continue their troubled life of ash, steam and boiling water eruptions. From Ruapehu to the Bay of Plenty vigorous geothermal fields are common, some with their geysers and hot water streams used early by the Maoris as village sites, others now harnessed to electricity generators. The modern city of Rotorua uses the steam and hot water under its streets and house sections as an abundant source of domestic power. The volcanoes sometimes kill, like Tarawera near Rotorua which erupted in 1880 to pour mud and fire over its namesake village and kill 153 people, and Ruapehu which erupted down the Whangaehu River on Christmas Eve 1953, and swept away a railway bridge in the path of a passenger express, killing 151.

Geologists don't know how the New Zealand crustal mass became concentrated some 600 million years ago. While the amount of land above the sea is small, the crustal plateau on which it rests covers a vast undersea area particularly to the south and east. Perhaps the mass broke off from continental Australia and drifted to its present position at the time when the old Gondwanaland mass fractured to release the Southern Hemisphere island continents? Perhaps the silaceous, granite-type rocks of New Zealand floated to the surface of the molten basalts of the ocean floor independently of any other continent-building activity? Certainly the country's land mass has risen and fallen beneath the sea many times in its mountain building process. Great plains of sedimentary rocks have been laid down under the sea and have been lifted as the earth's skin wrinkled—up into the heights of the South Island's Southern Alps which rise to 4 000 metres in peaks like Mount Cook.

The Maori's own story of the beginnings of his earth is really not too inaccurate in some ways. In the beginning Rangi, the Sky Father lay upon Papa the Earth Mother and mated with her—the sea covering the drowned land. Their children forced the parents apart: particularly Tane, the life giver, fertilizer, and sustainer of all living things. Correctly, the first land was the South Island—geologically speaking its rocks are older than the North—and the hero demi-god Maui caught the North Island on his line when fishing and raised it above the waves. Tane clothed the earth with

The shape of ancient ice. The glacier-carved Cleddau River valley, near Milford Sound.

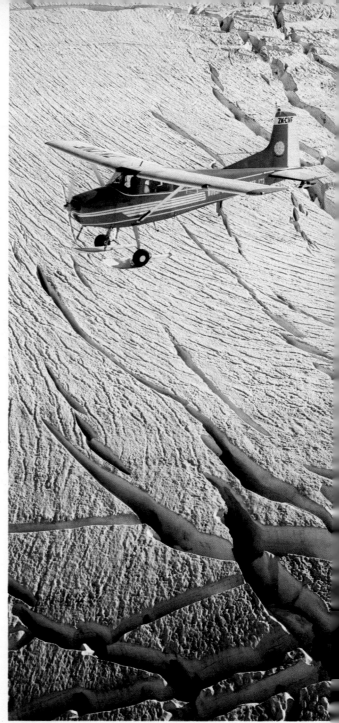

Above left:
Nor'wester coming at dawn, Canterbury.

Above:
Tourists fly over ice falls with cracks several metres deep.

Left:
Ice crystals on glass, Central Otago.

Opposite above:
Mount Cook (right) and the great dying glaciers which formed a landscape.

Far right:
Consider the lilies of the field.
A face to the sun on the stony ground and the high air. The Mount Cook Lily.

Opposite:
The Kea, a bird of almost human character, which haunts the high country.

forests, birds and other creatures, and the sky with sun, moon and stars. He had had to cut off his father's arms before he could throw him into the sky and the blood which flowed became red ochre in the earth.

The Maori also said, 'People die, are killed, migrate, disappear; not so the land which remains forever'. His land was owned communally and tribal boundaries were very clearly defined, broken down again into the bounded areas owned by *hapu* or sub-tribes. Possession of the land was maintained by use and occupation. The Maori had a passion for his land and said the proverb, 'By land and women are men taken'. Possession was by the moral force of *ahi ka*, the lit fire in the *whare* or house, the cooking pits of the *marae* or village. Essentially though, the Maori lived in a sense of what the philosopher Lévy-Bruhl called *participation mystique* which means the idea, universal among primitive peoples, that what happens outside themselves also happens inside themselves and vice versa. The Maori's creation myth allowed him to feel the processes of each earthly day—sunrise, the growth of plants, the death or dismemberment of animals—inside himself, and likewise to feel that his own thought and care for the processes of life and death could also influence them.

New Zealand seems to provide the possibility of this kind of involvement in processes in far greater measure than most countries because of its variety and the lushness of its range of physical environments all within a tiny space on the surface of the earth. Seeking experience, the New Zealander can learn what it might feel like 'to be in England now that April's there' by watching spring happen in Christchurch or Dunedin. He can comprehend the vastness of a desert by travelling to Central Otago or the bleak wastes of the North Island's volcanic plateau in winter time. He can understand what a tropical sea might be like if he lies in the bottom of a boat drifting among the mangrove inlets of Auckland's Waitemata Harbour; he can comprehend a tropical jungle if he strays into the drenched rain forests of the South Island's west coast; the alpine wastes of Tibet in the high country of the Southern Alps; the steppes of Central Asia in the MacKenzie Country of South Canterbury; the peasant farming of Indo-Chinese mountain country in the bleak villages of the Wanganui River and the vine and tree-fern-strangled ravines which streams cut deep into the soft papa mudstone of back-country Taranaki; the silence of the great Canadian pine woods in the half-million acre, man-made forest which covers the chill and barren plateau of Kaiangaroa. He can know the exquisite seas of the Greek Islands by diving in the crystal waters of Te Kaha, and the cool of the Italian lakes at Queenstown on Lake Wakatipu.

I once worked on a road mending gang near Dunedin, my home town in the South Island, and our foreman was an old soldier from World War I who rode with the Otago Mounted Rifles Regiment attached to the Australian Light Horse as General Allenby fought the bloody Turk through to free Jerusalem. One day we poured concrete for the surround of a new storm water drain at the road

Maori children watching a rugby match at Ngaruawahia.

Left:

A big wind from the west has left sunrise cloud over the lonely beaches of Lake Taupo. Named from *Taupo-nui-a-Tia*, 'the feathered shoulder cloak of the great Tia'. Tia, the lake's discoverer, had a long sleep on the night of the day that he found it.

Above top:

Pohutukawas live for centuries. People camp under them.

Above:

Winter in the old country of Piriaka, King Country.

Opposite:

Grown to a great size in peace and crystal-cold water, the rainbow trout of Rainbow Springs, Rotorua.

Annual cattle muster, Molesworth Station, North Canterbury.
New Zealand's own Wild West, a continental-sized cattle range in a South Pacific Island. Rehabilitation of rabbit-ravaged land by Government farmers has brought life back to 186 000 hectares. There are over 150 thousand beasts.

side. Next day, as we lowered the heavy iron cover into place old Jim patted each corner in turn.

'There we are,' he said. 'The four corners of the earth.'

He belonged to the earth, and as a New Zealander he felt that it was his to grasp and shape. New Zealand had prepared him for it.

When J. B. Priestley was in his eightieth year he visited New Zealand and in Dunedin told me how he had felt, as he travelled up and down the country, the presence of some animus in the landscape, some primeval force that was hiding itself from him behind a mountain range, or the dense edge of a podocarp forest. I told him that I had written in an earlier essay, 'The bush stands silent, waiting to come back'. Priestley said he felt there was some force 'out there' trying to persuade him not to paint the landscape, something indifferent but determined to preserve its privacy. He may have misconstrued its intentions.

R. A. K. Mason, the great New Zealand poet of the 1920s, 1930s and 1940s, wrote about the animus of the land as well as his own love in a poem called 'Flow At Full Moon', which embraces at once the concepts of the secret forces of the bush and mountains, of the land being devoured by the sea as its rivers carry their burden of mountain sands across the plains, of the irresistible tides of life and season which so affect man here.

Beloved your love is poured to enchant all the land
   the great bull falls still the opossum turns from his chatter
   and the thin nervous cats pause and the strong oak trees stand
    entranced and the gum's restless bark-strip is stilled from its clatter

The scope of mystery is more concrete, however. While the herds of red and other deer have continued to flourish in the bush and high country, the Canadian moose, also introduced, has vanished in the fog and rain-drenched, moss-hung jungles of Fiordland. In spite of attempts to find survivors of the stock originally released, expedition after expedition has found only 'sign' of the beast, never a wary animal. In a few hundred square kilometres large creatures have vanished because the country remains so wild that it is virtually unexplored. As recently as twenty years ago, sizeable lakes were being discovered for the first time in this most remote corner of the land, and ground expeditions were being sent in to examine forest and mountain areas unaffected yet by either man or introduced animals.

It is not surprising that people still talk of finding alive one of the great moa, the giant flightless rail that used to roam freely over the country but probably were nearing extinction three or four centuries ago. New Zealand is a land of birds—except for a small and shy native rat, a rare native bat, and a now-extinct dog brought by the Maori, New Zealand was almost entirely without land mammals

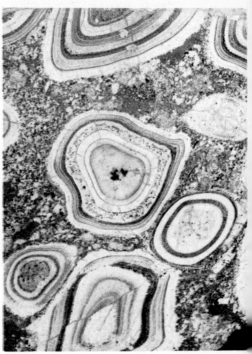

Above left:
The Cleddau River valley, near Milford Sound.

Above top:
Sunrise on a mountain stream, Southern Alps.

Above:
Granite spheroids made in the great heat of the old earth.

Left:
Aurora australis, over Ashburton.

Opposite:
A pattern of living. Ploughed field, Canterbury.

until the first release of the domestic pig by European visitors in the late eighteenth century. Pigs soon developed a vast feral population in both islands and provided a staple meat diet for both Maori and early European settlers alike.

The land of birds was a land of strange birds. As well as the giant moa of the southern plains there was a giant eagle which preyed upon them. There is still the flightless kiwi, a shy night bird with a long curved beak to dig for forest worms, and the flightless weka or woodhen. A flightless parrot, the kakapo, a lethargic-seeming green night bird living on grubs it pulls from rotten wood, is almost extinct. The remote valleys of the South Island's glacial lake, Te Anau were in 1948 discovered to be the last home of the flightless takahe, or notornis, a bird thought to have become extinct before the turn of the century. Today it lives on, precariously preyed upon by introduced stoats, ferrets and brown rats. A dozen or more other bird species from tiny bush wrens to the blue-wattled crow of the North Island rain forests face extinction from the same predators. The regal huia whose white tail quills, set among the burnished black of its body feathers, were worn by Maori chiefs to show their rank, has also vanished. The sacred white heron, kotuku, whose name means 'once in a lifetime', breeds in one small swamp area of Westland and will probably survive but its winter visitations are regarded anywhere as mysterious. Three white herons it is sometimes said, carry the souls of the dead to Te Reinga, the place at the tip of the North Island where men go down into the spirit world.

New Zealanders have recently been put to the test on how well they treasure their heritage of bird-life which is extremely profuse for such a small land area. First there was a nationwide confrontation between conservationists and the Government on whether or not the level of Lake Manapouri should be raised to provide storage water for electricity generation to meet the needs of the Comalco aluminium smelter. Raising the level would have destroyed the lake's long shoreline so that unique wildlife habitats could never recover. A nationwide petition raised 400 000 votes against raising the lake and it was spared. More recently the New Zealand Forest Service announced plans to mill hundreds of thousands of hectares of native beech forest on the South Island's west coast and replace them with farms and man-planted forests of northern hemisphere pine tree species. The conservationists who regarded the primeval forests as an irreplaceable heritage were defeated, but for the first time all New Zealanders, three million of them, brown and white, became aware of a real dilemma in planning the future of their country's wild heritage.

Such dilemmas are natural. The people who live on New Zealand's 77 000 farms managing 61 000 000 sheep, 2 000 000 dairy cows, and 7 000 000 beef cattle, who grow 97 130 hectares of wheat, 526 100 hectares of other cash crops and specialized fodder, have literally *made* the soil which produces green grass and other plants in such abundance. They have fed and pampered the soils, using

Whitebaiter at Hokitika River where gold dust flowed in the black beach sand.

Below:
Maori with melons, Waikato.

Below:
And even in the land of birds and trees there is a desert. Sand dunes, Ninety Mile Beach.

Above:
Hokianga Harbour, where the first Australian settlers sailed to obtain timber for masts.

Below:
A sea jewel in the wet sand.
Paua, the New Zealand abalone.

every mechanical device known to agriculture and a good number more of their own invention like specially designed hay-making machinery, or aeroplanes for spreading superphosphate and grass and clover seed on steep hill country faces. In their preoccupation with changing the land to make it productive they had lost sight, until recent years, of the need to preserve its special and original excellence for growing forest trees, native grasslands and a bird and insect fauna unique in the world.

But the population density of the whole country has so far reached only 28 people per square mile. Otago for instance, one small province, has a land area greater than Denmark or Switzerland with a population density minute in comparison, close to the rural low of four. New Zealanders have space—not the wastelands or daunting emptinesses of Africa and Australia but a space that men can creatively fill, in which the rewards to the spirit accruing from understanding and developing the land are positive and immediate. And the rewards from protecting the land, as the European is suddenly learning to do in a belated act of understanding the ancient Maori's veneration of the primal forest, are great.

Even in the cities like Auckland, Wellington, Christchurch and Dunedin a new mood is developing that can countenance concepts of conservation. In the late 1930s and the 1940s successive socialist governments poured the people's energy into providing for themselves the nuts and bolts of existence—sound houses built on a piece of land large enough to give a man and his wife a sense of dignity; health services which would ensure that their children grew straight and strong; education services which would discover through their universal application the human resources previously buried in the morass of working class poverty. Workshops and factories were built and New Zealanders began to understand, dimly at first, that they could develop skills more intricate and demanding than digging post holes for farm fencing or labouring on the Public Works Department's roads. They began to see that they belonged to a nation which could compete in industrial skills with the European nations from which their country was born. They began to understand that they could survive apart from 'Mother Britain' and that her influence was stunting their growth. They began to see that the United States was not really a South Pacific or Southeast Asian power at all but an interloper, and that they as a Pacific people close to Asia, possessed the moral force of belief in their own identity within an international context of common interest.

New Zealanders have always been travellers—like Bob Buckley who lived in Antarctica working at Scott Base on the Ross Sea coast as an electrical engineer and later toured Europe building up trade in the opossum skins he trapped on his Palliser Bay farm. New Zealanders shift easily about the world, learn and return to their favoured islands. Many reach high places in human affairs. Most return to their country because it is alive, growing, and even yet unspoiled.

The brown, desert-like landscape of North Otago, south of the MacKenzie Country.

Opposite:

Tena koe, E hoa, greetings to you, my friend, Ngaruawahia

Above top:

Sunrise, Castlepoint lighthouse.

Above:

Tree ferns in the Coromandel Range.

In the early sixties I began a book about New Zealand with these words: 'Man is a shadow on the landscape. These hills and mountains, even the harsh and level plains, seem to shrug off human conquest. The bush stands silent, waiting to come back. The sea is waiting to flood in again; volcanoes sleep fitfully and may awake at any time; along the great earthquake zones which underlie cities, the mountains are waiting to split and swallow down multi-storeys of masonry, trolley buses, pigeons, pavements and roof-top sunbathers.

'In New Zealand there are no beasts of prey. Even in the sea, sharks take a life infrequently. One mildly poisonous species of spider is rarely seen. No snakes coil under their arid rocks, the carrion birds are small and silent, no poisonous fish infest the snow-fed rivers. Man was not expected here; He had to conceive imaginary hazards.

'And so there are stories of the *taniwha*, the mythical beast which haunts river gorges devouring the unwary; there are malign and Dionysian fairies, the *patupaiarehe*; there are hunted men who become legends because they learn to hide in the forests and mountains; there are *ponaturi*, night-time invaders from the depths of the sea.

'Throughout a few million years of being plunged beneath the sea, or covered with glacial ice, of the inundation of volcanic rock with its hot rain of ochre ash, of the grazing huge flightless birds, and of the fire of man, the bush has learned to re-assert itself. Soft air and sunshine, damp winds from the surrounding sea, make it an easy life for plants, whether they grew in the original silence or came in the ships of settlers.

'These are islands. Perhaps no islands could seem so huge when you are imprisoned among their mountains and yet in flight above them you can see the circumscribing ocean on either side. No place is more than 130 kilometres from the sea.'

It is now possible to see how deeply the people of New Zealand have changed in the past twenty years. Except that, 'people die, are killed, migrate, disappear; not so the land, which remains for ever'.

Right:
The happiness of being young, free and a New Zealander.
Perhaps more than any other people, New Zealanders know how to make the most of their leisure time.

Far right:
From Norway's icy fjords comes a Norwegian fisherman to Whangaroa Harbour, Northland.

Below:
The ubiquitous trampers.
Young people with packs on backs are a common sight throughout the country. Many are Australian students 'seeing New Zealand on the cheap' during their vacations.

Opposite below:
Bus, prams, mothers, New Plymouth.
The New Zealand style of carrying prams on a rack over the suburban bus's front bumper. The driver is efficient, impatient to get going again.

Above:

The shores of Whangarei harbour where, according to Maori legend, the jagged rocks on the hilltop represent Manaia, his wife and two daughters and Paeko the slave who had seduced Manaia's wife. Manaia, becoming suspicious, had taken the family on a journey where a violent quarrel broke out between the husband and wife and her lover. Incensed with anger the *atua* (ancestral spirit) turned the group into stone on the hillside which is now named after Manaia, the jealous husband.

Left: Spectators.

Opposite:

Some use the land and sea in the old style, some make it new in the way humans do. Bethell's Beach, Auckland's West coast.

THE LANDBUILDERS

Late one summer morning in 1642 Abel Janszoon Tasman was called to his quarterdeck and told that the masthead lookout of his Dutch caravelle could see land. The ship was in the east of the Tasman Sea and Tasman took parchment and quill pen to write, for the ship's log, that he had seen 'a great land, uplifted high' to the east, and that he was sailing towards it. What he saw was the sheer western massif of the Southern Alps with its glaciers running down into the dense beech forests, and the black sand shoreline rich with alluvial gold which would wait 200 years to be gleaned by Englishmen, Irishmen, Australians, Americans and Chinese.

Tasman could not find a place to join the land and so sailed north into Tasman Bay where his ship was attacked by Maoris in their war canoes. He had to have some of them shot to protect his crew. He left New Zealand soon after and on his return to the Dutch East Indian capital of Batavia, now Jakarta, drew up a map which showed a corner of the northern South Island and the southern North Island with nothing between or beyond. The land was, he thought, probably the edge of another great southern continent, the *Terra Incognita* of the South Pacific, which the English navigator, James Cook would attempt to discover more than a century later.

Captain Cook, in full command of his brilliance as an explorer, was to sail right round New Zealand in 1769 and 1770, charting its coast with quite amazing accuracy, and recording information on the country which had to suffice until organized settlement under the British Crown began in 1839. It was however Australia that Britain became interested in as a result of Cook's discoveries, and the beginnings of colonization—rather the beginnings of exploitation of New Zealand—had to await the arrival of the first convict transports in New South Wales and the setting up of a military and penal settlement there, before entrepreneurs began to consider the other country's riches. Seals, whales, spar and shipbuilding timber, and a native 'flax' which yielded a linen-like fibre, were the produce sought, and the first New Zealand white settlers were a sealing gang which camped on the shores of Dusky Bay in Fiordland for ten months in 1792. They had been left by a Captain Raven

Yarding, North Canterbury.

Corriedale sheep, a New Zealand 'invention' bred from English and Merino bloodlines at the turn of the century, rich in wool, fat in meat, now a genetic export round the world.

Above:
Sir Edmund Hillary, sometime Northland beekeeper, conqueror of Mount Everest. The face that reached 'the roof of the world'.

Left:
Pet lambs, Amberley Show, where the lambs' white woolly socks are as clean as the girls' synthetics.

Opposite:
Feeding out, a family activity on this Canterbury high country run.

Drafting sheep, Ferintosh Station.
Khaki shorts fade, wear, take patches like seventies kids' blue jeans in the city. And the runholder wears long hair and a greasy wool jersey.

who had a licence from the East India Company—curious now, to think that the company felt itself impowered to give it—to take fur seals on the south-west coast of New Zealand and sell them to China. Raven's convict transport, the *Britannia*, was chartered by some Botany Bay military officers to sail to South Africa and bring them the kinds of food they were accustomed to, as well as livestock like horses and cattle. He dropped his men at Dusky Bay on the same voyage and when he collected them, reported that they had lived well, without any illness, and feasted on fish and native water fowl and woodhens so healthily that their lack of bread brought them no harm. Their catch of 4 500 seals was considered disappointing but it was only the prelude to the wholesale slaughter of the Southern fur seal carried out by traders of several nations during the next thirty years.

New Zealand, therefore, was brought slowly within the ambit of British influence and a kind of imperial responsibility for the behaviour of white men in the northern parts of the North Island was hesitantly adopted by the Governors of the New South Wales penal colony. The basis in international law for adoption of such authority was dubious but was never tested. It did not need to be because Englishmen were taking the Bible to the natives of New Zealand, and English lives were put at risk. Apart from the ruffian white men—sailors, whalers, rum runners and assorted traders in timber, flax, pigs, tomahawks, glass beads, blankets and muskets—the Bay of Islands, destined to be the site of New Zealand's first capital city at Russell, was colonized by the Church Missionary Society.

The Society's conquest of New Zealand was led by the Rev. Samuel Marsden, a strange man who had a passion for owning land as great as his passion for religion, and became known in the New South Wales colony as a self-aggrandizing landowner as well as a particularly harsh Magistrate. If convicts were not willing converts, the Maoris of New Zealand at least had the virtue of having open minds about Christianity and it was Marsden who persuaded the Church Missionary Society to establish its New Zealand mission. On Christmas Day 1814 he held the first Christian service in New Zealand. As the years passed he made repeated visits to New Zealand combining trading pursuits and proselytizing with

Opposite:

Picking kutai (*mussels*) at Kare Kare, Auckland's West Coast.

Left:

Blue clay sluiced away by green water for pale gold. Scars of hydraulic-lift mining, St Bathans, Central Otago.

Bottom:

Rangiriri Tavern, South Auckland. The 'bush pub' of the early days becomes a 'waterhole' by a main highway.

Right:

Hydro-electric construction workers, Twizel. The strange act of vibrating concrete into a firm bed. Laying masonry.

Below:

'Life is for living, in love and full-grown.

exploration which did much to open up the north part of the North Island for later white settlement.

By the late 1830s the first thorough-going missionary effort, that of the Rev. Henry Williams, was well established in the Bay of Islands area and in 1838 Charles Darwin, the young scientist sailing with the H.M.S. *Beagle* expedition, was suprised to find how 'civilized' conditions on the Williams estates were, and was impressed by the transplantation of a little bit of England into the New Zealand wilderness.

On the other hand, he found the Maori to suffer badly in comparison with the Tahitian Polynesian. 'He may perhaps be superior in energy but in every other respect his character is of a much lower order. One glance at their respective expressions [convinces] that one is a savage, the other a civilized man. Both in their persons and their homes [the Maoris] are filthily dirty and offensive.' Darwin unfortunately saw only Maoris who had been subjected to the European's corrupting influence.

The missionary pioneers adopted a life style for settlers in the north that was to become a common pattern. They ate wild pork, birds, and fish. For vegetables they had kumara, maize and sour thistle and in hard times fern root meal, an ancient staple food of the Maori. Whether or not the missionaries, as they spread out across the North Island, laid up a store of resentment against the white man is hard to know, but like all white settlers they proved to be as concerned to own land as any moneyed Englishman coming to farm. Darwin had seen the 'savage' as hardly noble in Rousseau's sense, hardly the natural gentleman. There is no doubt that the Maori was generally either despised or patronized by the white man and as one commentator remarked to a fellow white settler with modern-day perception, you could hardly blame a Maori for being annoyed when told to leave white company because he smelled bad. All too often, missionary zeal and patronizing good intentions resulted in exploitation of the Maori—from practising the belief that the best kind of 'white' education a Maori girl could be given was one that fitted her for domestic service in a European farmhouse, to thieving the Maori's land outright.

The small settler was probably just as much a victim of European rapacity in this context as was the Maori tribesman. The settler embarked on a ship in England in three different classes—as a steerage passenger who could look forward to four months of bleak discomfort, as a 'second-cabin' passenger, or as a 'cabin' passenger who was given hotel-style service. All except the latter had to fend for themselves using the ship's company ration scale and galley cooking facilities. Thousands of settlers in the early years arrived in New Zealand knowing that they had been allotted a block of land but not knowing anything about what it would be like— it could be 40 hectares of steep *papa* mudstone gully and tree fern with no sun and an 2 000 mm rainfall, or 16 hectares of stony ground or swamp supposed to be tilled for grain cropping, or 5 hectares in a town settlement.

The pace of settlement after the Treaty of Waitangi in 1840 in which the Maoris throughout the land theoretically surrendered their sovereignty to the rule of law and Queen Victoria's Government, was

Maori freezing-worker, Canterbury.
Freezing-workers perform one of New Zealand's most vital jobs—processing the lamb and beef which are sent to export markets all over the world.

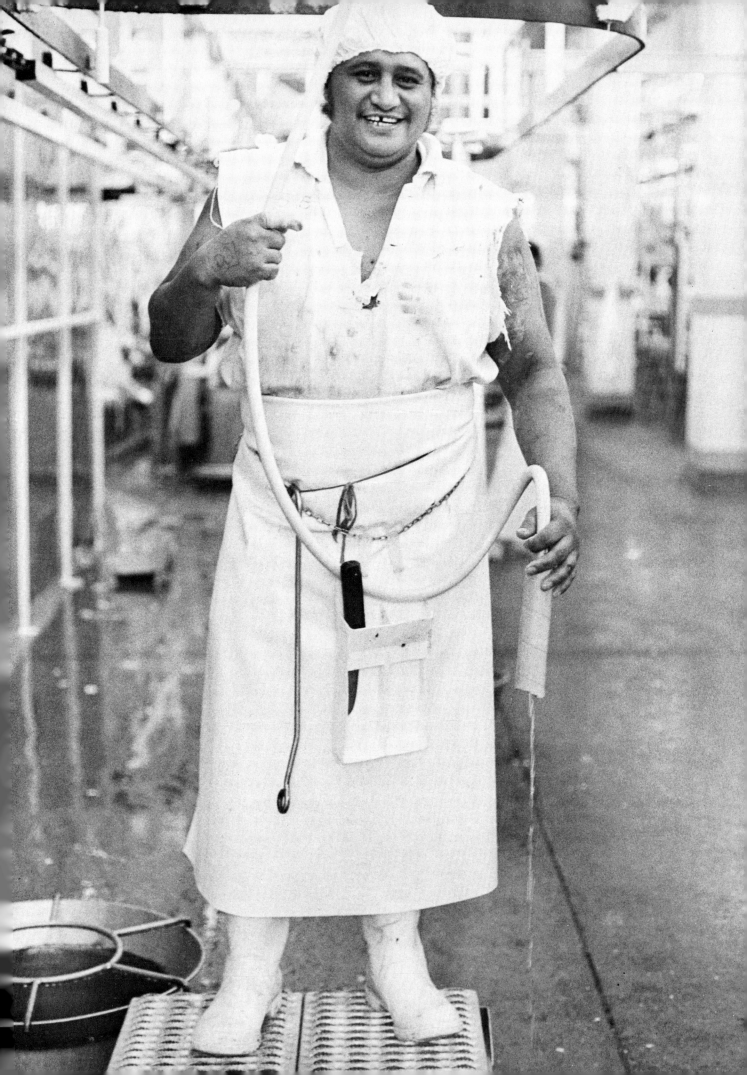

Right:
Like Patagonia. A 'gaucho-style' deer hunter carries his kill in the Kaimanawas.

Opposite and below:
Oldest wooden store building in New Zealand, Te Aute.
For more than a century the kauri timbers have served pioneers, tribal Maoris, farmers, Te Aute College schoolboys and travellers.

Saleyards, Eltham.
Men looked at beasts in this way when Robert Bakewell started breeding new stock in England's eighteenth century agricultural revolution.

so rapid that a real land hunger developed. Settlement, moreover, was managed by associations and companies like the New Zealand Company, which bought land in New Zealand, then sold it to immigrants in England along with passages out in company-owned or chartered ships. The architect of the Company was the politician and literary man, Edward Gibbon Wakefield, who imposed his own theories of social organization and political economy on the settlements he planned from the serenity of early-Victorian England. Broadly speaking, each settlement such as those at Canterbury, Nelson, Wellington, and New Plymouth was to be provided with citizens from a cross-section of English society so that a stable social system, serviced by capitalist land owners at the top, through professional men, shop keepers, artisans, down to common labourers could be established. It was not suprising that this system quickly broke down in the colonial situation since pioneering needed men and women with a vast range of virtues rather than specialists.

Perceptual geography is a new study with methods which can be applied in revealing ways to many of the New Zealand settlements. So far only the Otago settlement has come under its scrutiny—an investigation of the kind of place the first settlers of 1848 thought they were going to and the reality that faced them on arrival. Otago has a colder climate, too cold even for a Palliser Bay lemon tree and yet the first settlers, members of the Free Church of Scotland, believed they could set up a kind of theocracy in the new land which would offer social opportunities like land and education to their children, as well as freedom of worship. They also believed they were voyaging to a settlement with a Mediterranean climate. The shock of reality was such that the dissonance between their aspirations and reality continues to affect the life of their city and the effects of similar shock can be seen in others.

In the same way, the settlers of the North Island bush in the fifties and sixties were shocked to discover that their land was virtually covered in sub-tropical rain forest with dense vegetation and giant forest trees which had to be cleared before they could build even a shingle-board cottage, let alone buy stock for grazing and seed for sowing. Invariably, like the English at Horowhenua, the Bohemians at Puhoi, the Scandinavians at Dannevirke and Norsewood, the Lutheran Germans of the Moutere Hills, the settlers reacted spontaneously to clear the land, milling the timber, cutting and burning the scrub and sowing their seeds in the still-warm ashes. Sometimes the lonely homesteader would come home from a long ride to 'town' or a desperate day's work in the next valley to find his house a smoking ruin and his wife and children tomahawked in the kitchen garden. In the battle for the land he was as much a victim as the Maori whose land he was taking. He was settling it without malice, as he saw it, and moreover, according

Left:
Milking is over in a Northland cow shed.
Northland dairy farmers gave New Zealand its first alternative political party in decades. They voted in a Social Credit M.P. in the early sixties.

Above:
To keep the threshing dust out of the eyes. A wheat farmer harvests his crop.

Below:
Maori men, Ngaruawahia wood chop.
Arms that build nations.

Opposite:
West Coast miner at home on the banks of a sluiced gully.
Not much more than buttercup gold comes to an old prospector's dish these days. Once the stream held nuggets to fill a fist.

to a contract he'd signed in good faith in faraway England.

The greatest success of the Wakefield plan settlements was really their lack of success in doing what they were planned to do. Nelson, for instance, was one of the most 'pure' theoretical establishments but the inbuilt pomposity of the system which the English tried to impose on the Nelson area was quickly shown up by the Maoris who massacred a party of twenty-three whites beside Marlborough's Wairau River in 1843. The English had been out to teach the brown man a lesson for what they would have called 'insubordination'. The Maoris led by the Chiefs Te Rauparaha and Te Rangihaeata, made the only logical response available to them. The Wakefield settlement at Wellington also quickly broke down because there were not sufficient goods and services available to maintain it. In the first year of the settlement labour leaders campaigned for and won the basic right of the eight-hour working day. This was the work of Samuel Duncan Parnell, a carpenter who called a meeting at Barrett's Hotel—still a well-known Wellington watering place—and persuaded his mates that 'anyone offending should be dunked in the harbour'. New Zealand was the first country in the world to work to this rule.

In other parts the New Zealand Company philosophy of settlement did see the establishment of a kind of 'squirearchy'. These were New Zealand grazing lands, the vast areas of native tussock grasslands, particularly in the South Island, which could support hundreds of thousands of sheep.

Here came the gentlemen sheep farmers from Victoria and New

Opposite:
A Maori farmer from Te Araroa, on the fringe of the Urewera. During World War II he served as a sergeant in the Maori Battalion.

Below:
The camaraderie of wool buyers. Buyers from many nations compete at wool sales.

Below left:
A witty artist has decorated this North Island grain silo.

Above:

Deer hunters, Kaimanawas.
The yellow horse with the Arab nose, faithful pack animal and high country helpmate, always led, never leading.

Above left:

The 'faller', Rimu Forest, Lake Ianthe, Central Westland.
A petrol-driven power saw replaces a Sheffield-made axe but the face of the woodsman and the bush remain the same.

Opposite:

Kinloch Homestead, a mansion on a large sheep farm at Little River.

Right:

'Mark of the Lion', Charles Upham, V.C. and Bar, war hero, one of an extraordinary colonial family.

South Wales as well as the young squires from England, and in the grazing lands of the South Island the first legendary white men appeared, like McKenzie, the Gaelic shepherd who spoke no English. With the help of an amazing dog he is said to have stolen 1 000 sheep from Levels Station, South Canterbury, in 1856 and driven them to Southland, crossing mountains and snowfed rivers and giving his name to a vast sub-alpine basin. Today he has become subject to the kind of recognition that the bushranger Ned Kelly is still accorded in Australia.

In 1861 gold in Arabian Nights measure was discovered in Otago. Tens of thousands of miners came to New Zealand from New South Wales and California and when the rush was over, spread out over the land to provide the labouring force without which agriculture could not have developed, nor roads and railways have spread out from the cities to the farming centres. In ten years from 1861, £13 000 000 worth of gold left Otago and on its way established Dunedin as the country's industrial and financial capital as well as the birthplace of ideas which would make New Zealand one of the world's first Welfare States.

The goldfields were places where democracy ruled because no other political system but one of sharing could survive. Likewise the later men who dug for amber in the Northland kauri gum fields, or coal among the rich rocks of the South Island's West Coast, the men who sailed the ships of a country dependent on sea transport for its economic survival, the men who drove the railway tunnels, and spanned the rivers with their bridges, the men who learned to slaughter sheep in the vast abattoir works and freeze them for the world's first frozen meat trade from the southern to the northern hemisphere, the men who sheared the wool and harvested the grain in their short seasons of useful work, all became convinced that the human heritage was one in which men *were* equal, and in which every man's contribution was unique.

As one of the first Wellington settlers wrote home to England, 'No tyranny rules in this blessed land: when it pleases us we can take our gun or our rod in our hand, and stroll along the shores and in the woods, and shoot as much fowl and take as much fish as we like; no taxes are levied to oppress us. We may say with Tell—"Blow on ye winds, This is the land of liberty".'

The classlessness in the old sense of a day's work from every man, of the new society was described by an English bachelor: 'Fustian coats and thick shoes are very fashionable and you would laugh to see officers, doctors and dandies—digging, thatching and chopping with great frenzy... economy is the order of the day and I carpenterise, and carry logs, and cook, and go to council without detriment to my gentility.' So the tenth-hectare plots of the New Zealand town settlements were civilized.

Motorbike shepherd.
Men used to say they were happier on a horse but an 'ag bike' is easy riding.

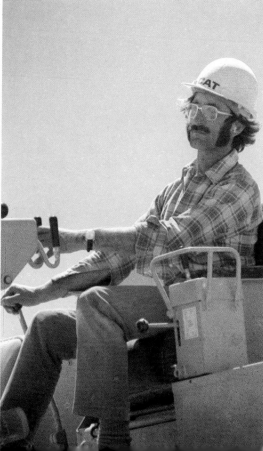

...bove:
...pdressing superphosphate. An animal act ...m an aerial machine.

... left:
...iry farmer, Hamilton.
...e cow cocky's rolled up woollen balaclava ...lmet hat, a symbol as apt as the fern and ... tiki.

... left centre:
...ozer driver, Waitaki basin hydro-electric ...wer construction.

...ft:
...ople grow like their environment. Hair ...outs from a forestry nursery worker, ...leith.

...posite:
...ize crop and silo, Cambridge.
...grow, is to hide, and be part of a burgeon-... earth.

THE MIXTURE

Thomas Kendall, the lay missionary, New Zealand's first schoolmaster and grammarian of the Maori language, used to say that, 'the true character of the New Zealanders is not so despicable as Europeans are apt to imagine'. He was to become rather more influenced by the Maoris than by his Church and Anglo-Saxon inheritance. In fact he 'went native' in the best tradition of British colonial pioneers. His comment pre-dated Darwin's observation and fell in direct agreement with the first impressions of later settlers. One New Zealand Company immigrant to Wellington, for instance, observed on arrival that 'the natives are exceedingly well disposed . . . they are perfect models of the human species and really are a splendid and superior race. They are intelligent, generous, faithful, open, and brave, and they will not brook an insult; they are honest, very honest, and will, if you treat them properly, do you many little favours'.

The year before, in 1840, a young Edinburgh doctor, John Logan Campbell first sailed up the Waitemata Harbour and landed on the site of present-day Auckland, wishing to buy land from the Maori. He called the place 'the Isthmus of Corinth of the Antipodes' when he described its beauties. 'The tui with his grand rich note, made the wood musical; the great, fat, stupid pigeon cooed down upon you almost within reach nor took the trouble to fly away. There was nothing to *run* away from us; for Nature, however prodigal in other respects, had not been so in vouchsafing any four-footed game . . . so poor *Tongata Maori* had to fall back on himself when the craving for animal food seized him and thus it may perhaps be inferred that land squabbles had oftimes a *belli*cose origin in more senses than one and that the organ of destructiveness was called upon to administer to alimentiveness, and cannibal feasts were the result.'

As a recorder of the original Auckland scene, Campbell remained quite unperturbed by thoughts of the Maori's potential violence towards the European although he understood all too well the effects that the European settlement would have on the Maori. 'For here [the Waitemata and New Zealand as a whole], there is no undiscovered country to which he can retreat and hold his own. He is face to face with that civilisation to which he succumbs. He is enclosed within a limited area, with a sea-board penetrated with innumerable harbours, with a fertile soil, with a climate the most genial the world knows, and by its speedy occupation he will be crowded out. For this land of which I write is destined to be the happy pleasure-ground of all the Great South Lands of the Pacific.'

Purepo Wik

Opposite:

Blowing it up at a beer festival.
Oompah beery band at a Bavarian Festival, Tuakau south-west Auckland. Descendants of 1840s pioneers.

Above:

'Fine piece of horseflesh.' Horse show judge, Canterbury.

Above right:

Chief Te Temekarepa, of Marlborough, at Waitangi.
Decorated with the plumes of the *huia*, the throat bell feathers of the *tui*, a chief remembers his past on New Zealand Day.

Right:

Harvesting, Darfield, Canterbury.
Under the brutal sun of Canterbury's flat country harvesters can dream of the beer to brew from barley they cut.

The Maori did retreat as far as he could go. The first signs of his suspicion of the settlers' intentions—being pursued as they were under the auspices of the Treaty of Waitangi which the chiefs made with Queen *Wikitoria* herself—came in 1845-46 when one of the powerful chiefs of Northland, Hone Heke, became jealous of the white power, and its symbol, the British flag flying at Waitangi, and chopped down the flagpole. Two years of skirmish warfare followed but the Maori resistance had not yet crystallized around the subject of land. It began to do so in the late 1850s when for the first time the white population of the North Island came level with the Maori population and the settlers' clamour for land forced their Government into legislative and administrative acts which both took the land from the Maori and attempted to break down his system of communal land ownership. The Europeans wanted to buy from individuals and under the communal ownership system adhered to by the sellers this was impossible. The white man re-inforced his demand for land by bringing in his professional army—the famed Redcoats — and thus the North Island was precipitated into twenty years of sporadic warfare. When the Maori resisted the taking of his land he could be said to be 'disobeying the law'. This was very convenient because his land could then be confiscated as punishment.

The wars ended—and their strange tales of heroism and chivalry which were later to create the myth of a just war waged by Europeans who, as it were, killed the misguided natives for their own good—were over by 1881. The Maori race was nearly finished too, dispirited in its disinheritance, corrupted by the European's vices, diseases, even habits of clothing, which were inimical to the Maori life-style and brought with them environmental ailments such as tuberculosis.

As he signed the treaty of Waitangi Captain Hobson, the first Governor-General, had said, '*He iwi kotahi tatou*', or 'We are one people'. In 1867 Maori men were given the franchise, unencumbered by property qualifications, twelve years before the white man received his rights.

But it was clear that being one people meant being European—assimilation or the death of a race in the terms that John Logan Campbell foresaw. The Maori went into a deep retreat. Generations earlier a dying chief had said 'Shadowed behind the tattooed face, a stranger stands, he who owns the earth and he is white.'

Above left:
Highland piper. After generations of New Zealandhood, the Scot's piper's moustache still nearly matches the red 'tooree on his bonnet'.
Above right:
Cooking for Italian tunnel workers, Tongariro power scheme.
Pasta brings a new tradition to colonial cookery. Food for a new kind of pioneer, the skilled European underground worker.
Opposite:
Bavarian beer festival, Tuakau.
After 130 years of settlement, a tradition still lives.

Opposite:
Master of the Hounds, Canterbury.
No foxes—but hares for the hounds to hunt, prize bloodstock to ride, and the traditional dress of another country's sport.

Right:
Brother Bernard, cellerman, Mission Vineyard, Taradale.
Mother Church runs one of the oldest vineyards. Wines vinted by monks have a proud place in a fast-growing wine industry.

Below:
Rodeo rider, Rewewhakaitu.
'Hey boy, you fell off, eh.'

Opposite below:
Roadside toy seller outside New Plymouth.
Do children really like fluorescent nylon fur? Fluffy animals are international.

Opposite:
The new mobility. A Polynesian girl hitch-hikes in a land where drivers are friendly.
Below:
Beaching a war canoe, Ngaruawahia.
The fugleman chants a work song for a great *waka taua* hauling out of the Waikato River, the most war-torn New Zealand waterway.

Fortunately, some Maoris learned how to play the game the European way, became lawyers, doctors, scholars, and politicians, formed the Young Maori Party and took up political roles. At the same time Liberal politics asserted themselves in Government, driven by the proletarian force of men and women who began to form an industrial working class in the New Zealand cities and demand rights like the universal franchise, factory legislation, a minimum wage, a pension system, and welfare services which people could enjoy by right rather than as charitable dependents of capitalists, merchants, and employers. The revolution in agriculture caused by high primary product prices following the Depression of the eighties and nineties, by improved agricultural implements, refrigeration and the milking machine, the establishment of adequate roads, railways and shipping services, meant that a steady drift of the rural labour force to the towns developed and the Maori, after a few decades, became part of it. Campbell was proved wrong. The Maori birth rate, instead of declining, began to rise rapidly. In 1896 there were 42 113 Maoris. Today there are 250 000 in a total population of 3 million. In the first years of European contact an estimated 80 000 Maoris had been lost by unnatural causes.

The Maori's economic and social decline after the wars, his loss of pride, helped to form many of the quite basic attitudes of prejudice and contempt and patronism which coloured the European attitude until the early 1960s. While they lived, no white New Zealander could afford to be patronizing about Sir Peter Buck, Sir Maui Pomare or Sir Apirana Ngata. He could however say, as a former Prime Minister J. R. Marshall did not so long ago, that 'Maoris do have a role to play in New Zealand society and I have been through factories and noticed they make particularly competent and happy machine operators'.

He could also be like Sir James Wattie, the millionaire vegetable canner, who endowed a prize for the New Zealand Book of the Year. In 1973 *Tangi*, a first novel by the Maori writer, Witi Ihimaera, was placed second and at the award-giving luncheon Sir James called the novelist to receive his due. On being told that many of his race worked in the tycoon's canneries, and on then being asked to 'rub noses' with him in an ancient Maori form of greeting Ihimaera could only ask, 'Is your nose clean?' a remark which made headlines in the Press.

In the past two decades the Maoris have re-assumed a self-confidence which most white New Zealanders had forgotten or never learned they possessed. There is now a strong Maori movement, particularly in Auckland which has the highest urban Maori population, towards the idea of 'two cultures, one nation' rather than the 'two races, one people', notion to which white men have given lip service for so long.

Groups like the Nga Tamatoa Council, the Maori Theatre Trust, and the Polynesian Panthers of Auckland's urban 'Polynesian ghetto' areas, are acting assertively in their campaign to insure that the Maori culture survives the experience of being swallowed up in white society through the too benign administration of welfare legislation intended to help brown people. There is even a white

Opposite below:

Assembly Hall, Te Aute College.

In these halls the Young Maori Party was born. Led by Sir Peter Buck, Sir Apirana Ngata and Sir Maui Pomare, it gave political focus to a nation and reshaped a people's destiny.

Opposite above:

A Canadian and a Welshman join New Zealand foresters in preparing land for a new crop of trees.

Right:

Norwegians run this factory at Norsewood which specializes in Scandinavian knitwear.

Below:

Emptying eels traps, Lake Ellesmere.

It took 100 years for European New Zealanders to understand the value of an eel fishery that Maoris have always managed expertly.

Below right:

Weeks ago a log hauler dragged the trees, which made the paper, which made the newspaper, which the log hauler crew reads at 'smoko'.

backlash typified by the decision of a recent Governor-General to discontinue a system of vice-regal National Youth Awards because youth itself insisted on nominating the Polynesian Panther leaders as worthy recipients. The justification of such a backlash is the claim that the militant policies of groups like the Panthers are fomenting racial feeling where previously there was none and creating a situation of conflict which owes more to the mis-directed energies of bored urban youths and their 'gangs' than to the inter-racial realities of 'our lovely country'. The era of debate which has been initiated is sure to have its continuing anxieties.

It is true that there are 30 000 Polynesians—Cook Island Maoris, Niueans, Tongans, Samoans and Tokaalauans living in downtown Auckland and bleak suburbs of worker housing like Otara. Maori leaders and social workers, appalled at the neglect of their special needs in education in the New Zealand language, law, custom, food, and basic codes of behaviour are beginning to have their cries of 'black ghettos in Auckland' listened to. The South Pacific islanders who have migrated either temporarily or permanently to the acknowledged South Pacific capital city are part of a movement of peoples the Government for many years felt powerless to inhibit. Recently, however, immigration officials have been required to trace illegal immigrants and return them to their homes. The Island people want education, money, status, and they believe that they can get it in New Zealand and take it home with them or export it to relations left at home. In many cases they are forced into migration to earn money to support relatives left struggling in Island economies and social systems which have been destroyed by contact with Europeans and the influence of their civilization.

Through radio, television, and the newspapers New Zealanders are becoming endlessly self-critical about these problems, without demonstrating much real adaptability in reaching multi-racial harmony. When newspapers start publishing 'employment offered' ads in three or four Pacific Island languages and when supermarket hoardings present shopping information in the same way to aid the bewildered Polynesian housewife, the white New Zealander is inclined to feel that something immoral is happening and that whatever it is is not his fault. When activist Maori groups campaign for Maori to be taught in New Zealand schools as a second language, the middle-aged New Zealander of today remembers how he was taught 'Maori myths and legends' in primary school and wonders what all the fuss is about. His inability to form opinions contrary to his agreeable memories of Maori 'fairy tales'—for those were what he was taught in forms Europeanized and stripped of their inner religious meanings—makes him feel that militant Maoris and their white supporters are 'just stirrers' with nothing better to do. He remembers how his taxes have helped pay for a special Maori school system, for lower mortgage rates for Maori housing; how nobody used to care that Maoris were not allowed to tour Africa as members of All Black rugby teams and that leaders of Maori rugby have been openly in favour of continuing to play 'racist sport' with South Africans. Perhaps he remembers how his grandmother used to 'sew for the Maoris' on Wednesday afternoons

Above:
Saturday morning chores, suburban Auckland.
An Auckland advertising executive in hat bedecked with beer can tear-tops, works out with a lawn mower on a spring morning.

Above right:
A combination of grace and concentration.
Perhaps one little girl in five thousand realizes her childhood dream of becoming a ballerina.

Opposite:
A cutlet of hapuka, (groper) **from a Greek fork, Christchurch.**
Oysters back-to-front in Greek-style letters. For New Zealanders, a new enthusiastic way of selling fish.

Above:

Wiha Malcolm of the Waikato.

Left:

Anglican Vicar, St Paul's Memorial Church, Putiki.

A church, decorated with the sacred skills of Maori art, stands on once bloody ground where Chief Te Rauparaha, with nearly 1000 warriors attacked Putiki *pa* on the Wanganui south bank.

Below left:

Watching life—an ageless human occupation.

Opposite:

Mrs Hapimana, of unknown age.

One of the last New Zealand women with the chin *moko*, carved in the skin with a chisel made from an albatross wing bone.

Far left:
Sunday morning, Samoan Presbyterian church, Ponsonby, Auckland.
For expatriate Samoans seeking the consumer society and money for families at home, Christianity is solace for homesickness.
Opposite:
Nicola Nobilo, Auckland wine grower.
Passionately a New Zealander, a Yugoslav grows wine, owns an Irish setter and builds a house of stone like the ancient family buildings of the European homeland.
Below:
Watching the country sports.

at the Ladies' Guild and how annoyed she was at their ingratitude when they helped to vote in the first Labour Government on a social welfare platform in 1937 and how angry she was when she found that all four Maori electorates would continue to return Labour members of Parliament in the future. He begins to compare the Maoris unfavourably with the 6 000 Indians, the humble greengrocers and shop keepers who are New Zealand citizens and rarely upset anybody, or with the 10 000 Chinese New Zealanders, survivors of the gold rushes, who have proved their 'assimilation' by their hundred years of good behaviour and are now regularly joining the professional classes.

The fact is that assimilation—in this context, absorption by the white society of the Maori society—has not worked. The Maori has seen the white man as acquisitive, success-seeking, cold, and isolated in his nuclear family. The white man has seen the Maori as spendthrift, unreliable, pleasure-seeking and immoral in the relaxed manners of his extended family.

The latest visit of *Kuini Eripata* — Queen Elizabeth — to New Zealand was memorable for the Maori placards saying 'You Are Not Our Queen. You Are An Imposter' and the attempts to bomb indignity into the celebration of New Zealand Day, formerly Waitangi Day, the anniversary of the Treaty's signature. White governments around the world have traditionally failed to realize that the only cure for racial disaffection is to spend vast amounts of money on assisting the minority race and putting up, with as much dignity as they can honestly master, with its lack of gratitude. Maoris have had rather less to be grateful for than the European thinks and he still typically protests that giving the brown man lots of money will be no solution because it will 'make him soft, he won't work, he'll go back to the mat'.

New Zealanders now ought to be in a potentially more hopeful position because they can look at the effects a massive application of welfare state policy has brought to their country as a whole and recognize the fact that their society has only gained greatly increased vitality from its political experiments. At any rate, white New Zealanders can be sure that the Maori renaissance of the 1970s will not end in the disarray of the 1870s.

Left:
Sons of another earth. A Dutch farmer's children, Tokoroa.

Below:
Dutch Bar, Dutch barman, Hastings.
Delft china, pale and bitter-sweet Dutch lager, French stained glass, provide a place like home for Dutch New Zealanders.

Opposite:
Bathing in the rain in a Rotorua hot pool.
The subject for a Maori Renoir or Gaugin. The geothermal pool is hot enough to make the warm rain feel icy.

THE TOWNS AND CITIES

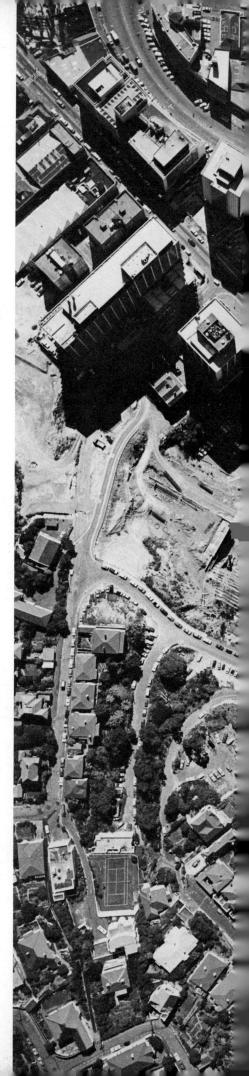

As the land of New Zealand was subjected to its wars, migrations and disappearances of people, the towns began to take on their first kinds of stability. Probably no single invention of the nineteenth century did more to stimulate the growth of urban centres and the development of a new kind of New Zealander—an urbanite—than refrigeration and Britain's concurrent creation of huge fleets of refrigerated shipping.

Before the first shipment of refrigerated meat from Dunedin in 1882, there was no way of sending meat, butter and cheese out of the country on their long voyage to northern hemisphere markets. In the old days sheep at the end of their productive lives were killed and their carcasses boiled down to make tallow which could be exported. Wool was shorn, sold and bought, on the farm. Likewise the 'stock and station agent', the man who bought produce and sold the farmer seed, the implements of farming, his monthly grocery order, and auctioned his stock for market to meet the needs of other farmers, was a travelling man working between isolated homesteads.

Refrigeration meant freezing works and dairy factories and towns to hold pools of labour to serve them. It meant sophistication of the transport industry to accommodate vastly increased production, the rapid growth of a Civil Service to administer the laws of production and commerce, the establishment of importers of goods and of agents in the distributive industries, of transport workers like the 'wharfie' as stevedores became known, of the growth of engineering to service the farms and the transport industries.

Before 1882 the New Zealand agricultural labourer had not been highly skilled, trained or educated in a general sense. The refrigeration revolution meant that he had to learn to use his brains as well as his brawn to apply technical skills when he came to the towns and cities, and so more urban people were required to teach him in schools and train him in industries.

In Auckland, Wellington, Christchurch and Dunedin, the centres which would become New Zealand's first true cities, the old social order was radically changed in quite a short time. The first great upheaval of the working class movement came soon after

Wellington, a city built on an earthquake-tortured landscape.

Opposite:

Sunset over Waitemata Harbour and the Auckland Harbour Bridge.

'For this land of which I write is destined to be the happy pleasure ground of all the Great South Lands of the Pacific.' John Logan Campbell, on first seeing Waitemata Harbour, 1840.

Below:

Auckland's Vulcan Lane, the air of Oceania's Capital.

Braless girls, their mates in platform shoes. Vulcan Lane, with its craft and music shops, boutiques, bistros, old as settlement and part of the sweet variety of a youthful metropolis.

refrigeration—in 1889 when a coalition of coal miners, seamen, and dispossessed rural workers suffering from a recession in agricultural produce prices went on strike for better wages and conditions. The strike was necessarily 'broken' after several weeks but at the next elections in 1890 a Liberal Government was brought to power and the first Labour members elected in coalition with it.

For the Wellington 'establishment', the society of moneyed people engaged in importing and commerce, banking, the law, medicine, the shock was considerable. One Wellington paper reported in its social columns later that year that 'Sassiety has been ruffled' by a report that a Wellington 'lady' had been asked by her maid if she might see the morning newspaper. When asked why on earth she wanted the newspaper the maid replied that she wished to read an account of her brother's maiden speech in the House of Representatives to which he had just been elected as a Labour Member.

Those Fabian socialist sophisticates, Sidney and Beatrice Webb, visited New Zealand in 1898 to see for themselves the operation of the world's first Industrial Conciliation and Arbitration Act, and found 'sassiety' very boring—far more so in Auckland than in Wellington, the home of Parliament and the administration. They reported that politicians and other social leaders were hardly elevated and idealistic in their thinking. Rather they were pragmatists, men of common stock who debated grand affairs of state as if Parliament were a union meeting hall or a country auction sale. New Zealand was dull and the average wage of three pounds a week earned by urban workers was sufficient to keep them well fed and clothed and incurious about the higher things of life.

The Webbs continued to keep an eye on New Zealand, to see it go through the convulsions of great periods of labour unrest in 1912 and 1913 in which the 'landed gentry', as they now considered themselves, of the Wellington, Auckland, and Christchurch provincial areas, enrolled themselves as special constables and rode into town to help bludgeon sense into the striking urban workers. 'If blood be the price of your cursed wealth by God we have bought it fair', the unionist banners read and they were to go on buying it for another twenty-five years until the old freedoms of Wellington's Petone Beach in 1840 began to be restored by the first Labour Government.

Part of the enigma of the New Zealand urban person has always been his desire to emulate the establishment figures in his milieu and his intense loyalty to the British Crown and what it represents in New Zealand. Today his cities reflect his primary aspirations for a better material future. He wishes to compete not through 'improving himself' in 'sassiety' but through owning more and better things than his neighbour and keeping their appearance better. For the past fifty years 'the quarter-acre section' has been the goal of urban people in general—a concept of substantiality which goes back to the ethos of the first settlements—and has been achieved by a remarkably large number of people. At its worst level the ideal has resulted in estates of government-built housing which are barren, soulless and depressing to extremes, like Otara in Auckland or Wellington's satellite city of Porirua. In such places there is an air of people having been offered their Promised Land,

Opposite:
A young city-dweller.
The main cities have much to offer the young at heart who flock from quiet provincial towns to share flats in Auckland, Christchurch and Wellington.
Far right:
The City of Auckland Administration Building symbolises the confidence and dominance of the largest city.
Below:
'Now there was a painter.' Looking at a touring Constable exhibition, Auckland Art Gallery.

Above:

The metropolitan explosion. Guiding a crane sky-high in downtown Auckland.

A workman's hands define the old and the new, a pioneer's church framed between them, modern man's steel and concrete looming into the future.

Opposite:

Outer reaches of Waitemata Harbour from Remuera.

'The brooding presence of Rangitoto'—a tourist book description of the old Gulf island volcano, but in a slate-grey sky and sea it becomes part of the city.

Below:

Student motor cycles, Auckland University.

Under the plane trees of Auckland University's campus-side streets students park the emblems of a new social mobility.

Opposite:

State housing, railroads, giant factories, Lower Hutt.

A Welfare State dream—Government-built housing crowds into industrial territory in the Hutt Valley. The box houses and long-run roofed factories are dwarfed by the raw hills.

Below left:

The 'wizard' who entertains those from seven to seventy in the Christchurch Cathedral Square.

Below:

The Wine and Food Society, a recent phenomenon in New Zealand's cities.

The remarkable improvement in the quality of New Zealand wines, together with a new appreciation of local foods has encouraged pride in a national cuisine. For too long the excellence of New Zealand delicacies—lamb, venison, whitebait, smoked eel, scallops, oysters—was neglected in favour of the ubiquitous international menu of 'onion soup, pepper steak and chocolate mousse'.

entered it, and felt trapped in dormitory areas. The result is quite contrary to the idealistic welfare state motivation behind the construction of such new towns.

There is a new mood in New Zealand's cities. The fruits of the welfare state that was so despised in the fifties and sixties by that part of the electorate committed to private enterprise and laissez-faire economics—that is, most of the rural land-owning population, the distributors, and the traders in imported consumer goods—have been an emerging class of young professional architects, teachers, publishers, lawyers and industrial managers who, through their education, have come to see a new purpose for the city. They have been applying their social and political influence to reverse the process of stagnation taking place in the down-town areas of the country's major cities. For instance the Wellington Architectural Centre, strongly influenced by brilliant young architects like Ian Athfield, Christopher Brook-White, Roger Walker and Keith Wilson, has changed the course of mid-city development by persuading public bodies like the City Council and the Harbour Board to preserve the natural aesthetics of the city environment.

The Wellington group has 'arrived' as powerful people in the Age of the Environment, but just too late to stop public works like the Wellington motorway system which will eventually destroy much of the central city's character with its huge tunnels, bridges, over and under-passes that lead nowhere but to a city's heart already suffocated by motor vehicles. In Auckland too, groups of radical architects from the Auckland University's School of Architecture, are campaigning for the re-development of down-town Queen Street, the big city's main artery, to preserve its historic character and sense of dignity and space under the South Pacific sun.

In Dunedin architects like Ted McCoy are helping the University of Otago to plan and design a major new university campus integrated with the city's beauty, which will accommodate several of the special schools of the New Zealand tertiary education system. It is growing from the complex of land and buildings which made up New Zealand's original university. In Christchurch civic leaders, with the help of architects like Peter Bevan and Miles Warren have embarked on an urban renewal programme led by the architectural successes of their new Town Hall and the complex of sports stadiums, swimming pools and future university campus accommodation specially built for the spectacularly successful 1974 British Commonwealth Games.

The other new mood of the New Zealand cities is the mood of consumption. Education and industrial development have produced a new kind of New Zealander—members of the managerial class. Typically they live in garden suburb sub-divisions on incomes just sufficient to support their standard of affluence. As well as mortgaging themselves to their homes and motor cars they are similarly tied to their carpets, their colour television sets, their second car, their trailer-born motor boats, their golf clubs and their patio barbecues. Their sections are no longer quarter-acre. New Zealand is not a big country in terms of usable land and the most far-sighted among architects, town planners and agricultural economists are continually

Opposite:
Wellington City grows out from its centre of settlement.
Cold Southern Ocean currents welling up in Cook Strait bring an Antarctic clarity of air to some Wellington mornings.

Below:
The new Wellington Club, The Terrace, Wellington.
The gentlemen's club of a nation's capital. In the classic colonial wooden building which preceded the new Wellington Club, much of a nation's history was written by the powerful financiers and politicians of the nineteenth century.

pleading for the better use of land already under housing instead of continuing the destruction of prime soils for growing the country's greatest resource, to accommodate new sub-divisions.

The urban New Zealander sees his progress towards bigger and better houses as part of the reward for striving for bigger and better jobs, and his household shift is always one step further towards that outer-suburban retirement-home settlement with its better quality houses and gardens, its quieter life, its ultimate total privacy. Feeling adventurous, vital, in a hurry, he seeks to replace the pioneering of his commercial life with the pioneering of his great-grandparents whose first battle was with the bush.

But in a real sense, and though he seeks it so ardently, the New Zealander has failed to come to terms with loneliness and isolation. The nuclear family has been his downfall and he reacts in two different and paradoxical ways. As centres of his community social life he builds enormous taverns which can only be called drinking barns, thousands of square metres large and with many thousands of square metres more outside for car parks. He defies his loneliness by carving up the most beautiful seaside, river and lake-side areas in his land into somewhat less than one-tenth hectare plots, half-size replicas of his suburban homes and gardens, where he goes to 'relax' at weekends. The purpose of the 'bach at the beach' is not to create for himself a richer life of social contact. It is to be a retreat from life itself. The huge drinking barn is just such a retreat, for in such places man is anonymous. He can stay isolated, get drunk, drive home in his personal car.

New Zealand cities and suburban areas need the corner pub just as they need the corner store which habitually breaks Shops and Offices Act regulations and stays open all weekend. The affluent society needs to be 'humanized'. It needs the development of localized community feeling. Such community feelings do survive in country areas. Small country towns tend to be kind and tolerant

Opposite:
Tokoroa, a new town built for a new forest industry.

Out of sour volcanic soil grow good grass, forest trees, houses and healthy babies. The New Zealand Forest Products-planned housing development of a timber town.

Opposite below:
A summertime audience amuses itself, waiting for music and dance. Cushion Concert, Auckland Arts Festival.

Below:
Little brown jugs, Saturday afternoon tavern.
'You're telling a lie!'
'It's the truth, man, she really was.'

Far left:

Opening of the 1974 Commonwealth Games, Queen Elizabeth II Park, Christchurch.

Silence over one of the world's great sports stadiums. The clapping wings of pigeons herald the opening of an international festival, January 24, 1974 and 70 000 people wait for the unique celebration of human prowess about to begin.

Left:

Christchurch Cathedral, midday in autumn.

Trees draw down sap from their autumn branches, people come to the peace of a cathedral square where cars are banned.

Below:

The Ferrier Fountain, Christchurch Town Hall.

Christchurch hits world class in the architecture of its public buildings. The Warren and Mahoney-designed wonder of New Zealand's new idealism in city planning.

Bottom:

City shopping, time for laughter.

It's a free country and it's lunchtime in Christchurch's Cathedral Square.

places, timeless, now that the pioneering is done, in the rhythm of rural life. People tend to accept each other's potential for living a happy life and rarely question each other's deficiencies.

New Zealand's major cities also exist in a state of extraordinary jealousy towards one another. Not long ago a journalist wrote a story about how Aucklanders enjoyed seeing an industrial fair-exhibit extolling the virtues of Dunedin and Otago, as a fine free place for New Zealanders to live in. A Dunedin paper, bitterly aware of the vast development of Auckland compared to its town's economic stagnation, billboarded the story, 'Drift to South Starting'.

On television Auckland's famous and eccentric Mayor, Sir Dove-Meyer Robinson—known throughout the country as 'Robbie' retorted that Aucklanders' interest in Dunedin was merely northern politeness. He didn't see any reason why the 'drift to the north'— the migration of New Zealanders to their largest city and the best urban climate—should stop. 'Coming to Auckland is like kissing a pretty girl,' he said. 'You can't stop.'

Above:
One of New Zealand's greatest assets, its youth.

Opposite:
A champion of the unborn child's right to life, Professor Albert Liley, CMG, Research Professor in Perinatal Physiology at Auckland University's Postgraduate School of Obstetrics and Gynaecology. His delivering hands save lives in the National Women's Hospital.

Left:

Hotel lounge bar, Cambridge.
Oldtimers mourn the departed long kauri bar the bare floor boards of the country town pub but the consumer society demands 'elegance'

Below left:

Stopping for the School Patrol, Hamilton.
Traffic education begins at kindergarten age for New Zealand children. Boys and girls control pedestrian crossings with the authority of policemen.

Below:

Suburban woman, Bell Block, New Plymouth
On land disputed in blood, a housewife tends her suburban paradise. On this soil the most bitter fighting of the Taranaki Maori Wars broke out in 1860.

Opposite:

Youth and the motorbike, Bluff Hill, Napier
Kids and power, petrol, concrete and paint Ingredients of an urban revolution still benign in a family-sized country.

THE ECONOMY

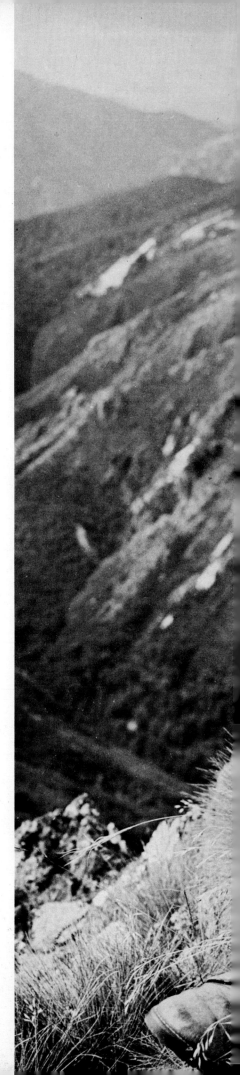

However uneasily the New Zealander lives with his welfare state he is not dismayed by it and he is finding creative ways to finish its structure. In the mid-seventies the economic resources are available in an abundance nobody ever really believed they could reach. In a country so deeply tied to Great Britain from 1770 until 1970, deep pessimism about New Zealand's chances of survival when Britain entered the European Common Market dictated the course of social and political change in the decade of the sixties. At earlier times there had been various attempts by New Zealanders to break away from Britain's 'apron strings' as they have been constantly called, but even the most radical political leaders who triumphantly pushed through the welfare state legislation of the 1930s and 40s were loath to adopt into their country's constitutional precedents the kind of legislation which would allow them to rule an economy free of British trading self-interest and British political influence.

The Statute of Westminster, a document drawn up at an Imperial Conference in 1929, did offer New Zealand freedom from the overriding power of the British Parliament to pass laws inhibiting New Zealand law or vetoing New Zealand attempts to establish a legal system more immediately adapted to her South Pacific situation and national identity. But the Statute was not ratified by New Zealand until 1947, at the end of the first Labour Government's term of office. From a contemporary standpoint the reasons, apart from anxiety about defence, are hard to understand. New Zealanders at last do have a sense of nationhood. Their continuing successes in international trade are opening up new market opportunities, not only for primary produce but for an increasing range of manufacturing goods. New part-processed products from growing assets like commercial forests have given New Zealanders proof that Britain with her century or more of trade imbalance has indeed been a 'paper tiger'.

Deer hunter, North Island high country.
Time out, and no doubt another $50 lying in the tussock to be taken out on the professional shooter's shoulders.

Above:
Building the gas-fired electric power station New Plymouth.
Symbolic sparks. Building the metal works which will help fire a nation with a new source of electrical energy—natural gas.

Left:
New production hole being drilled, Kapuni gas field, Taranaki.
Sweating on the rig. Lubricating mud covers the summer-hot drill platform workers who meet the challenge of a new technology.

Opposite:
Setting up a drilling rig, Kapuni, Taranaki.
Far at sea off Cape Egmont towering oil drilling rigs have discovered a natural gas field twenty times the size of land-locked Kapuni, a major field by world standards.

New Zealanders had to learn that they had a right to assert themselves in trade and in international politics. New Zealand politicians, radical or conservative, consistently adopted the attitude that Britain could not be offended in trade or foreign policy because Britain bought almost all of their country's wool, meat, butter and cheese. In return New Zealand bought manufactured goods from Britain. New Zealand remained too timid in this relationship to assert itself. In the depths of the Great Depression of the thirties successive New Zealand Governments failed to insist on tariff agreements with Great Britain which would allow the development of secondary industries in New Zealand, a matter of life and death if an urban society was going to survive in the country at all. British traders in fact, dictated tariff policy to New Zealand Governments and kept thousands of urban workers in New Zealand on the dole, in relief work, or in non-productive public works, until they were saved by the advent of World War II.

And so, still lacking the audacity to ratify the Statute of Westminster, the charter for freedom of development, the New Zealanders bent their backs in the post-Depression recovery to pay back development loans worth millions of pounds sterling which fell due in 1940–41. Part of the ethos behind such behaviour was the feeling that if New Zealand did not satisfy the conditions of Imperial Trade, then the Empire would not guard the country in time of war. At least that is what common people believed as they struggled under the fiscal disadvantages of satisfying Britain. As it had done in World War I, the country again gave in World War II thousands of its young men to a European war and suffered for the second time in a mere thirty years the destruction of a generation of talent, intelligence, inventiveness, curiosity and creative insight. Not content with this expiation, New Zealanders made Britain a gift of 10 000 000 pounds sterling in 1947 to help the Mother Country in its economic recovery, money which they already owed her in loan finance repayment anyway.

Now that the population explosion, a world food shortage and monetary inflation have insured New Zealand an ever-growing income for so long as humanity survives and needs meat and dairy products, the nation has reached a true economic independence. In a world of shortages in consumer goods New Zealanders have the skill and inventiveness to make their own. The days of constant political warfare between the farmers and the importers of manufactured goods on one side, and the coming New Zealand industrialists and manufacturers on the other are certainly over. Both sides will benefit from world-wide scarcities, and the ordinary New Zealander will benefit most.

Above left:
Making an Airtrainer CT-4, New Zealand Aerospace Industries, Hamilton.

Above right:
Forestry research, a vital part of a profitable and rapidly-developing pulp and paper industry.

Opposite:
Aluminium ingots smelted at Tiwai Point, Bluff.

Opposite right:
Built back into its forest shape—paper rolls from man-planted pine.

Above left:
The tallyman. Timber workers write on durable wood.

Above:
Slicing logs for transport loading, Kinleith forest.
Jaws of the 'grasshopper', as timber workers affectionately call the long-legged machines which move their logs.

Left:
Starting a 'controlled burn'.
Pioneers burned the bush to plant their seed crops in the warm ashes. So did Maoris to plant kumara. Now Maoris burn a seedbed for northern hemisphere trees.

Below left:
Paper mill at sunset, New Zealand Forest Products, Kinleith.
Like a Saturn booster rocket at Cape Canaveral, a giant tree digester gulps down wood to make paper pulp.

Below:
The Papermaker, kingpin of the burgeoning paper industry.

Right:
New Zealand Forest Products paper store, Kinleith.

Below:
A sculpture in a stack of logs.

It may be that some New Zealanders will feel sad that the battle for economic diversification with its attendent social benefits of a more vital, creative, and ebullient society with more sophisticated tastes, and an environment which can begin to throw up those individuals who will be the new and ever-renewing leaders of the human race, has been won by default. But they should not despair because at least the desire of two generations for economic and political independence has built a society with a heavy potential for creative leadership. It may now be time for civilization to be reborn in The Antipodes and for Australasia as a whole to become the word's most important incubator for new ideas.

New Zealand now has a huge aluminium refinery, steel produced from its own mills using iron-sand ore and with an ore surplus for export, vast reserves of coal which are now, in the time of world energy crisis, a vital asset, a major natural gas field in world terms off the coast of the North Island, and vast lodes of other minerals waiting for a technology to exploit them. At the same time it has a large proportion of its population determined to see that resource development is done according to the highest principles of environmental conservation. Britain, as it moves ever closer to the tight cartel of old European countries, is steadily becoming less meaningful to the New Zealand economy and its viability as a source of investment capital is also rapidly withering.

Opposite:
Dr Pat Holland, research scientist.
At Ruakura Animal Research Station, Hamilton, Dr Holland searches for ways to build better soils to grow richer grass.

Below:
Sheepmuster, Mount Linton Station, Southland.
Ten thousand cross-bred sheep, breeds that built a nation, come up for shearing to the yards of a vast South Island run.

Opposite below:
The Comalco Aluminium Smelters, Bluff.

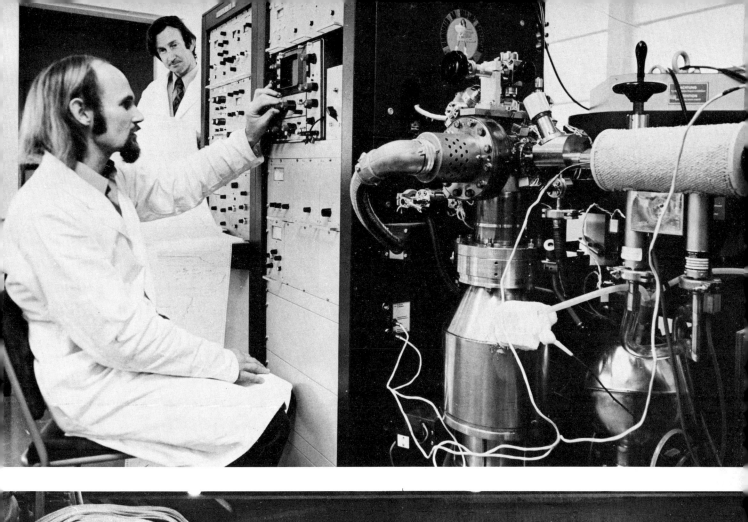

Left:
Furnace worker, Glenbrook.

Below left:
Pouring molten steel, New Zealand Steel mills, Glenbrook.

Steel from the sea. From the black ironsand beaches of the North Island's West Coast and electricity from mountain water and natural gas, New Zealand's first steel industry was born. The plant will produce 153 000 tonnes a year to meet basic industrial requirements.

Below:
Making penstocks for a hydro-electricity station, Christchurch.

Through vast steel tunnels will pour the stored energy of a great South Island alpine river and its lakes, through energy collection systems designed and made by New Zealanders.

Bottom:
Sheet steel rolled at New Zealand Steel mills, Glenbrook.

Opposite:
Steel pipes, New Zealand Steel mills, Glenbrook.

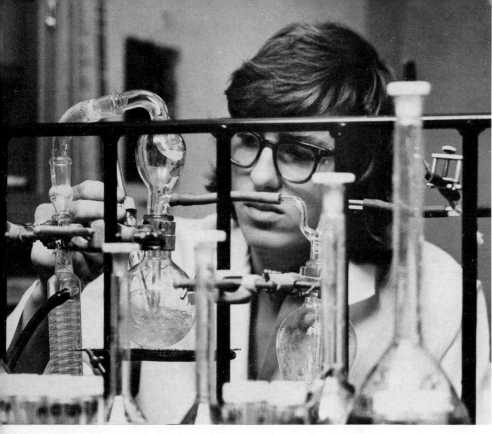

Early in 1969 the New Zealand Government put its people through the think-tank of what it called a National Development Conference. It tried to seek guidance from the people as a whole about how national production, consumption, social and environmental goals could be achieved. Notably the data on production and the conclusions drawn from it, were soon found wanting, but the New Zealander, at the Conference and since, did manage to tell his Government about something which it could achieve in a fairly measurable quantity, and by degrees which could be quite accurately planned. In effect, New Zealanders rejected the concept of having a planned economy with its careful restriction and modification of rewards for labour, along with its insistence on the the fulfilment of productivity targets. The Conference organizers had been careful to point out that their planning system was only indicative—a projection of future production and the planning of other economic vectors on the basis of assumed growth—and yet the hundreds of New Zealanders involved in the project would not accept even this form of economic control. Wage earner and businessman alike stuck to a kind of private enterprise concept in which the wage-earning unionist saw himself as a self-employed person with a commodity to offer in the market place in the same way as a home appliance manufacturer might offer pop-up toasters or refrigerators.

Above:
Pollution control laboratory, Comalco's New Zealand Aluminium Smelters, Bluff.
In the world's cleanest rich nation, ultra-pollution-conscious environmentalists insist that industry pay for using the air.

Opposite:
Power pylons march the length of the islands feeding industry.
Like birds on the wires, Electricity Department riggers set up new transmission lines.

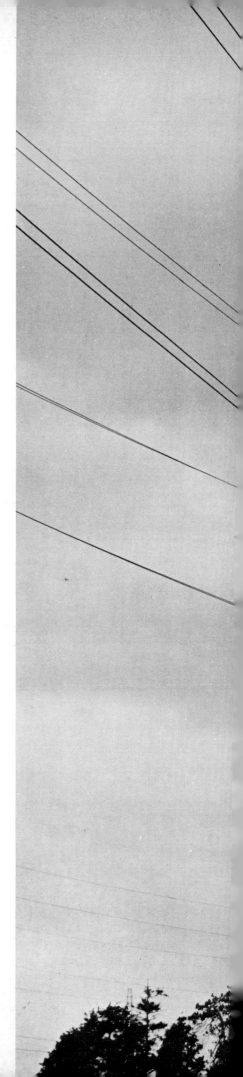

Left:

Eel processing plant, Lake Ellesmere.
Eels writhe and wait for the week-long wash in clean running water. Then they are electrocuted, frozen, smoked, sent around the world for consumption by gourmets.

Opposite:

Helicopter deer hunting, Kaimanawa Range.
A fat Red deer hind on her way to the restaurant tables of West Germany, one of an average 100 000 killed in a huge mechanized commercial shooting operation.

Below:

Oyster dredging, Foveaux Strait.
Export is forbidden. New Zealanders jealously eat all the 110 million oysters the stormy Foveaux Strait dredging grounds can produce in a year.

On the other hand, what was accepted, was the role of the State in welfare services. Conference delegates clamoured for State spending on improving 'the quality of life', a phrase only just coming into public debate at the time.

The New Zealander proved himself to be a thorough creature of his economy and his national life style. He indicated that a concept of government, freedom and political economy new to the world was emerging in New Zealand. Even the old tried-and-true system of Industrial Conciliation and Arbitration legislation broke down at this time, the hallowed Act of 1894 becoming obsolete almost overnight, and wage earners wholeheartedly opted for bargaining and contracting wage rates rather than accepting the measured dictates of a judicial body.

The New Zealander at last wanted to feel himself economically free, his own man, and in this same act of self-awareness he renewed and revitalized his concept of the role the State could play in fulfilling his other needs. He wanted Local Government to invest in cultural development, Central Government to spend more on recreational facilities, writers to be paid for their books borrowed in public libraries, actors to be paid professional wages for their work in regional theatres, more Central Government money for adult education in arts and crafts, and the freeing of public capital goods like school buildings for community purposes. He wanted liquor licensing legislation which allowed more 'civilized' drinking, a choice in his television service instead of a one-channel national system, smaller classes in schools, a thorough-going well-funded effort to introduce a tertiary education system in technology so

Opposite:
Dr Eugene Chen, maker of animal vaccines.
In the heart-land of the Waikato dairy industry, a Taiwanese veterinary bacteriologist devotes his skill to New Zealand livestock.

Below left:
Shopping in Cathedral Square, Christchurch.
The Dreams of men. Urban living brings new roles for members of a changing society.

Below right:
Beer tanker driver emerging from his tanker after a cleaning operation.
The largest of these modern tankers can carry up to 2 430 gallons of draught beer over long distances

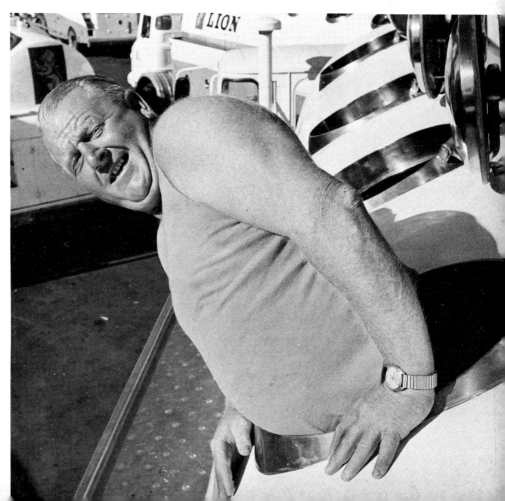

Opposite:

Clothing for a nation, woollen yarn at Kaiapoi mills.

Right:

Testing turbo-jet engine, National Airways Corporation, Christchurch.
Engineers work round the clock in their test laboratory.

Below:

'Airtrainers', New Zealand's export aeroplanes, New Plymouth airfield.
A New Zealand aeroplane has reached commercial production status and a large export market years before car-making is a profitable proposition.

that his talents could be increased and fulfilled. He wanted more facilities and trained people in his national parks, and a universal and free pre-school education system.

When he demanded these things he said that he wanted them free and un-encumbered, as of right, as he had been used to. He did not want a regimented and suffocating cradle-to-grave surveillance from his government. He wanted his due as a human being. He is now beginning to get it.

The crucial condition of it all was his insistence that his quality of life should be improved without any increase in bureaucratic control. New Zealanders generally do not realize that their welfare services and Central Government investment in their communities are administered with a degree of form-filling and revelation of private information which must be the most discreet in the world. It has recently been threatened by a proposal for a centralized and computerised 'people information system' potentially more sophisticated than anywhere in the Western world but no doubt this is a short-lived planning aberration. New Zealanders have always been critical of the numerical strength of their Public Service and have failed to realize how exceptionally competent it is. The promise meanwhile, is that the country's social development will increase while its bureaucratic controls diminish and the New Zealander's greatest talent may turn out to be a capacity to live an ordered free-thinking, and fruitful life without being told how to go about it.

If this is so, the new people who keep coming to New Zealand from the old northern hemisphere countries will find their choice justified. They will find open communities in which they can apply with dignity the skills which have made them eligible under the country's immigrant entrance qualifications. For immigrants, skill and good health are essential because, in spite of some belief that the country could open its doors to immigrant labour as Australia did after World War II and that the immigrants themselves would create the industrial employment they seek, successive governments have decided on a very closely-planned immigration policy. And at last immigration from Britain has become controlled by the imposition of an annual quota. Even if the old philosophy which said that the Maori would be assimilated into the white culture and that as a result there would be no racial tension has been proved wrong, governments insist that European immigrants should be ready to become New Zealanders, relinquishing their older heritage.

Fortunately the New Zealanders themselves have not forced on their new neighbours such harsh imperatives. English, Scots and Irishmen, Greeks, Dutchmen, Hungarians, Czechs, Poles and Austrians have felt sufficiently at home in New Zealand, in spite of some difficult times in 1950s, the period of most intense immigration. They make special contributions to New Zealand life and open up new paths for the flowering of the New Zealand culture.

Opposite:
Discharge pipes, Wairakei Geothermal Power Station.
Super-heated water from as deep as 1200 metres in the earth generates steam to drive electricity generator turbines in a world which uses volcanic heat in the service of man.

Below right:
The pride of the Air New Zealand fleet, the DC10.

Opposite:
A punter studies the form.
New Zealanders spend more than $180 000 000 a year through the State-controlled Totalisator Agency Board on off-course and on-course betting.

Above left:

Taking out the squashed ones.

'It must be Wattie's': New Zealand's best-known advertising slogan from a giant food-processing company which literally feeds a nation.

Left:

Bean cutters, frozen food factory, Hastings.

'Whole French beans'—a new sophistication in the frozen food industry to meet demands of a newly food-conscious people.

Above:

Shipping at Mount Maunganui.

From a small holiday village has grown New Zealand's biggest timber export port, Mount Maunganui, Tauranga, the transport centre for a vast new industrial and agricultural region.

Opposite:

Apple washing, New Zealand Apple and Pear Board store.

In the post-Depression, 'nation-building' era, apples, like milk, were given to school children to make them strong and rosy. Now they are washed for the domestic market.

THE SCIENCES

My favourite New Zealand scientist has always been Joe Ward, nicknamed 'Chinese Joe', who used to be the director of the Dominion Physics and Engineering Laboratory, a division of the New Zealand Department of Scientific and Industrial Research. He was called 'Chinese Joe', because he liked and would quote Confucian philosophy and had statuettes of the Buddha, and various Chinese artifacts in his spacious office at the Gracefield Laboratories near Lower Hutt. At DSIR policy meetings he was said to stare at people with 'Chinese eyes'—an inscrutable look which he would adopt while waiting for a colleague to see the sense of his argument.

Ward was educated in New Zealand, went to England to work in electrical engineering research, then came back to New Zealand to work for the DSIR in Wellington. His great contribution to New Zealand science, and to New Zealand as his homeland, was to guide the DPEL into kinds of research which would have application in New Zealand industry. Apart from the scientists he gathered around him—brilliant men from many parts of the world—his was a rather lonely voice in science administration. His Laboratory made several major technological discoveries which could have been applied immediately to industrial circumstances, but society as a whole proved unreceptive and so his proteges eventually left New Zealand. Joe Ward retired and built himself a harpsichord instead. His major thinking as a creative administrator of a national scientific research programme had been in the realm of 'feedback', that phenomenon which is basic to the whole concept of electronic industrial automation. Many of the discoveries his people made are still waiting to be put to use by industry, mainly because New Zealanders have been slow to provide money for their development, and slow even to provide organizations to put the new knowledge at the country's disposal.

Ward's work fell squarely in a long New Zealand tradition which had its origins in the curiosity of Englishmen trained in the natural science disciplines which arose from the concerns of the French encyclopedists of the late eighteenth century and flowered in such talents as Darwin's, Hooker's, Huxley's, and Forbes', in the life sciences and those of the great Royal Society geophysicists of 1820–1910. Traditionally also, such men worked for government

Advanced birth induction research by Dr R. A. S. Walsh, Ruakura.
In a land where a stud bull calf might found his own million dollar dairy industry enterprise, a cow gets sophisticated medicare.

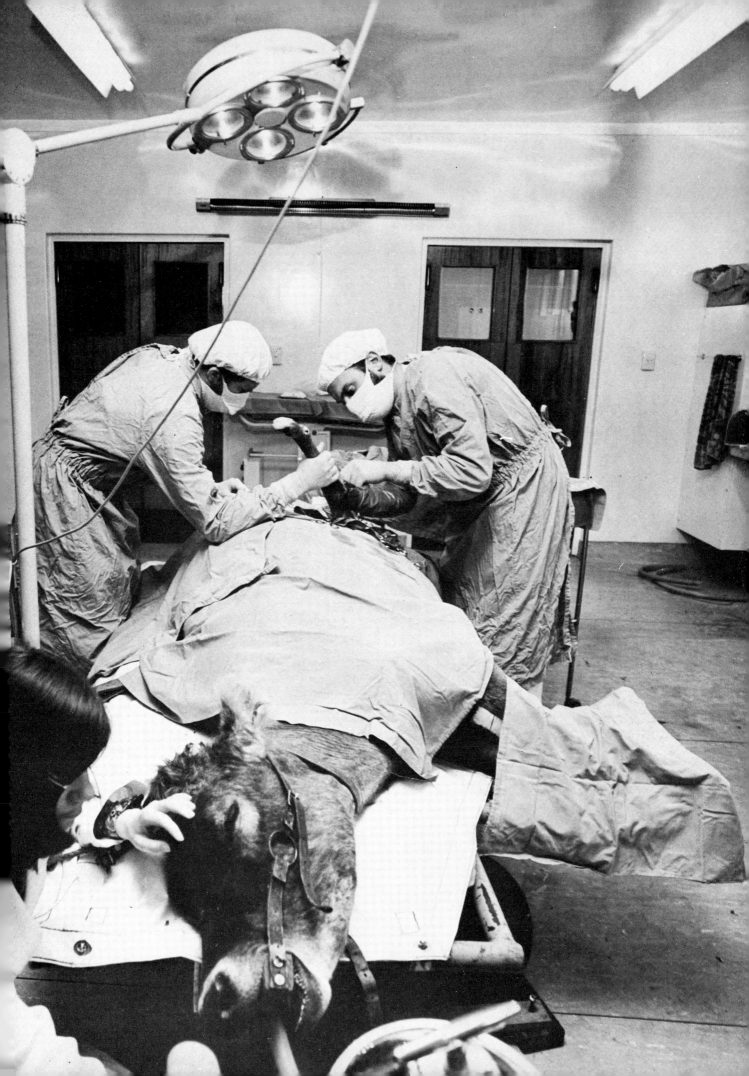

Above top:

Rainbow around the sun.

A strange sun on a clear day in the southern sky.

Above:

Bull semen stored frozen in liquid nitrogen.

In containers repeating the hexagon-structured molecules of life, the wealth of a nation waits in a frozen womb. The resources of the artificial insemination program of a huge dairy industry.

Opposite:

The Mount John space watch station, Mackenzie Country, Canterbury.

High on an ice-carved Mackenzie Country monolith a United States space navigation and surveillance outpost glows in the light of an alpine electrical storm.

departments—like Sir James Hector, the nineteenth century geologist and explorer who came to New Zealand in 1862 at the peak of his exploring career. He became the first Director of the Colonial Museum and its complex of scientific departments, and of the New Zealand Institute set up by the Government to further work in the sciences and the arts, and later superseded by the Royal Society of New Zealand. Or there was Thomas Kirk who completed an unparalleled account of the New Zealand flora and for a time was New Zealand's first Conservator of Forests. Me like these, and others who filled the science and medical faculties of the country's emerging university system, set the standards for a research and practical scientific achievement which for a small country became enviably high. And so there have been Sir Harold Gillies and Sir Archibald McIndoe, world famous plastic surgeons, and Sir Brian Barrett-Boyes, a pioneer of 'open-heart' surgery working at Auckland's Green Lane Hospital, or Sir Arthur Porritt, Sergeant-Surgeon to the Queen.

All these human and individual successes of New Zealanders are completely outstripped by those of the scientists working in agriculture. In the early years of this century New Zealand began to face a great crisis. Refrigeration and the milking machine had made it technically possible for enormous increases in stock numbers and suddenly the country as a whole was found to lack large enough resources.

The job of turning the country into a gigantic expertly-managed farm was achieved by an extraordinary effort of co-operation and combined skill. The grasslands scientists experimented until they knew which types of grass suited different areas of the country, then bred new strains to give the best growth rate and palatability. The crop research scientists did the same with fodder crops like turnips, rape, lucerne and chowmolia and at the same time invented new techniques for making ensilage—methods of storing green crops with added nutrient chemicals. This was needed to feed stock in the dry late summers and autumns which affect many prime dairying areas, and in sheep-growing areas with dry winters and little feed available during the sheep's late pregnancy. The entymologists introduced new species of insects which would help to control pasture insect pests. The botanists tackled the huge problem of breeding a strain of wheat of hard-milling quality which would grow readily on the deprived soils of the Canterbury Plains or in the wetter regions of Southland and South Otago. They had such success that New Zealand became, and still is, self-sufficient in wheat. The soil scientists analysed soils where grass would not grow well and discovered what chemicals the soil needed to make it fertile. Their most resounding success has been in the Bay of Plenty-Volcanic Plateau region of the North Island where a cobalt deficiency had left animals 'bush sick' since the time of first settlement.

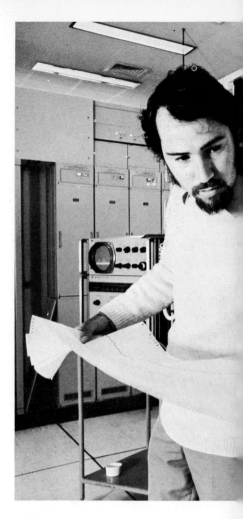

Above:

Post Office satellite communication receiver station, Warkworth.

Technicians check chart recorded information from internationally-owned stationary orbit communications satellites over the equatorial Pacific.

The stud breeders, helped by the veterinary and agricultural scientists, found ways to increase the fertility rate in sheep so that they produced more twin lambs, and discovered methods of artificial insemination of dairy cows so that the service of stud bulls from both New Zealand and overseas could be given to improve the dairy herds.

In its most spectacular expression, research joined up with industry and the individual farmer in the development of aerial top-dressing. More than 11 million hectares of New Zealand's 17 million hectares of farm land are too steep to be worked with tractor-drawn implements. In the mid-1940s pilots and aircraft engineers returned from World War II to pick up the threads of an idea which was just tackled in the 1930s. Using war surplus light aircraft they began to experiment with sowing wealth into the land in fertilizers. Gradually they developed techniques of flying manure into country previously thought to be impossibly steep. Then they began flying fencing wire and posts into the same areas. In the first few years New Zealand farmers bought sufficient wire to put a fence twice around the earth at the equator. In a decade the New Zealand farm mushroomed. When they had poured manure on the land, now in special aeroplanes they designed themselves and had had built by New Zealand's first aircraft engineering industry, the scientists envisaged another task. They developed strains of clover seed which could be spread from the air, then a method of coating each seed with sufficient manure to start its growth, then a method of innoculating it with nitrogen-forming bacteria as well. Not only did the clover grow in previously impossible country, the bacteria on its roots fed nitrogen into the soil and gave new heart to the old grasses.

But New Zealand's scientific effort as a whole remains lopsided while the fruits of industrial research are not capitalized. It was only in the late sixties that a New Zealand Standards Institute with adequate testing facilities, and a New Zealand Design Centre were set up. Meanwhile many of Joe Ward's men from the Dominion Physical Laboratory had moved on leaving behind a body of research and invention equalling in quality the work of the agricultural scientists. School children are invariably taught about that great New Zealander, Sir Ernest Rutherford, 'the man who split the atom', but rarely that in recent years New Zealand scientists have given them new methods of tape-recorded instruction for machine tool automation, new methods of moulding metal components, new applications of solar power for domestic use, new industrial uses for coal and the oily distillates of pine wood from their country's great forests, as well as revolutionary discoveries about how the earth behaves seismically, volcanically, and in response to the sun's energy which bombards it. Perhaps agricultural prosperity will now provide the time for them to learn.

Opposite:

Fisheries Research laboratory of the Department of Agriculture and Fisheries.

Algae bloom in laboratory bottles as fisheries research scientists seek ways to explore and exploit New Zealand's vast and relatively unknown fishing grounds.

Left:

Communications sateliite transceiving station, Warkworth.

Nightlight on the huge dish antenna of New Zealand's space age communications. The New Zealand Post Office is a part-owner of the world communications satellite system.

Below left:

Clouds of pollen drift through a spring pine forest.

In the southern land of trees, northern hemisphere pines grow so fast that tree farming has changed a nation's economy.

Below:

Cobalt sterilisation plant, Tasman Vaccine Laboratories, Hutt Valley.

The first industrial use of nuclear energy in New Zealand — a cobalt radiation source of sterilization for medical and hospital supplies, the first of its kind in the southern hemisphere.

THE ARTS

In the mid-1970s New Zealand is a glittering land. It is not just that the sunlight flashes on the sails of Auckland Harbour's myriad yachts, the towering new buildings of downtown Wellington, Canterbury's fields of ripening grain or Otago's towering quartz-gold mountains. There is a glitter in the people which is more than wealth and good food. It is a new kind of confidence, a new kind of understanding of life that embraces an idealism reaching beyond the old ideals of 'God, King and country', the negative aspirations for wealth and social exclusivity, into the wider affirmations of the human spirit. It is tempting to think that this may have come about through fifteen years of exposure to a national television system, the New Zealand Broadcasting Corporation, which consistently serviced its viewers with what it thought were the best quality television programmes available in the world. Through fifteen years of television New Zealanders have been given, for the first time as a mass audience (television ownership by households is at market saturation level) an objective standard on which to measure their progress compared with the rest of the human race. For the first time they have seen the work of their artists, musicians, and writers, while they all watched together and at once. They have measured their jurists, and doctors, and particularly their politicians. They have actually been able to compare what they have with what the rest of the world has, and have been asked what they think of themselves. The endless uncomfortable act of expatriatism which creative New Zealanders once had to make to find out what the rest of the world was like, and what real standards were, has at last become dispensable. There is an audience at home. The overseas journey will remain necessary but can now be made without that self-conscious diffidence or sheer awe-struck naivety which used to characterize the New Zealander 'abroad'.

The biggest success of the New Zealand Broadcasting Corporation was to create a climate which demanded its own dissolution, and the setting up of a new public broadcasting system which will give a wider choice and involve many more New Zealanders in the public creative processes of having their plays listened to, and paintings, music and performances subjected to an immediate critical response. The result may not reach expectations because the pool of public talent is small in most fields, but New Zealanders have

Frank Sargeson, novelist, playwright, extraordinary writer.

'Old maker of tragedies': the Mexican phrase for wise fiction writers was never more apt than for Frank Sargeson, his country's most enduring creative writer.

Above left:

International concert pianist, Michael Houston.
From a South Canterbury country town comes international musical genius. With second place in the Van Cliburn Contest, an assured world concert career began for a twenty-year-old.

Above:

New Zealand Ballet Company, His Majesty's, Auckland.
Every little girl's ambition—to be held aloft in ballet shoes and *tutu*, in the footlight's glow. Standards are set at annual competitions by local performing arts societies.

Left:

Dame (Edith) Ngaio Marsh of Christchurch.
More than twenty-five books in forty years, a unique woman of international publishing, theatre, entertainment, a leavener of her country's culture.

Opposite:

Music spans three generations in the Villa Maria Choir, Christchurch.

suddenly taken the cultural 'bit between the teeth'. They 'want to have a go'. Quite predictably it is with the same enthusiasm and intelligence that they have 'had a go' at warfare, rugby football, track and field athletics, sailing large yachts, horse breeding, and snooker, among other pastimes, and have been world-beaters at all of them. Now, having a go at television is just part of a general 'go' at culture.

For seventy-five years the expatriate New Zealand artists and writers left their country with a sense of despair. Their preoccupations almost invariably were with New Zealand and their experience of it. The burden of expatriatism made them think very deeply about the reasons why they had to leave and they began to define the land and its people with acute and often overwhelmed sensibilities. Until the mid-1950s few were able to express their feelings for their New Zealand experience in sufficiently universal terms to gain a world audience. Those who stayed behind probably found the task much harder but they stayed with it and laid the foundations for the cultural developments which are now taking place across the whole range of endeavour from television to op art and poetry.

One of the things which the expatriates fled from was what came to be called the 'down at the hall on Saturday night' mentality of most community-creative endeavour, the kind of tradition of country 'theatricals' which developed in the pioneering situation, the Country Women's Institute one-act play, the small town Repertory Society's attempt at Edwardian musical comedy or, if it dared, Gilbert and Sullivan. The Art Societies which have always flourished throughout the land have produced years of competent landscape painting but remained insensitive to the point of rudeness to adventurous artists breaking new ground in the genre of 'foreign schools'.

In terms of popular culture the real flowering of the New Zealand artist has been in popular music and entertainment. 'Talent quests' in the country district or on national television are now continually producing 'pop' and other singers of amazing accomplishment. Musicians flourish—from those members of the New Zealand Symphony Orchestra who formed in their spare time the New Zealand Jazz Ensemble with its thoroughly professional standards, to teenage boys and girls who sing professionally at anything from open air rock concerts to night clubs, and always with staggering poise and ability. They have created a new and universal kind of entertainment in a country in which 'family entertainment' has long been the tradition. Many now have international reputations.

The Queen Elizabeth Arts Council, the administrative body for public financial support to the arts, has begun to adopt a grass roots policy of fostering creative development while building on the social capital of the 'pop' arts and entertainments industry. Under a new chairman, the former trade and industry chief administrator, economist, social humanist and writer, Dr W. B. Sutch, the Council is trying to 'free' community capital invested in school buildings and church halls for community purposes. Its ally, quite suddenly in New Zealand, is an audience of people who have the money and will pay to see and hear the work of professional artists.

Opposite:
Maori rock drawings, Weka Pass, North Canterbury.
Many hundreds of years ago, Maoris sheltering under limestone cliff overhangs left typical examples of stone-age man's art.

Below left:
Rei Hamon, artist, in Coromandel pohutukawa roots.
A new talent with a new graphic art. Rei Hamon's exquisite ink drawings of living shapes in the forest have added another kind of culture-spanning dimension to New Zealand art.

Below right:
Potter at work, Taylorville, Westland.
A coal miner's hands stay deep in the earth. Hardy Browning turned to pottery when the coal face he worked at Taylorville was closed.

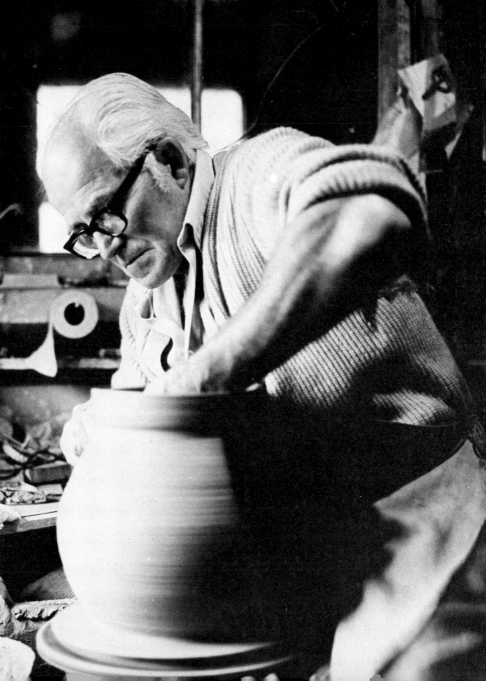

Above:

Pat Hanly, an artist whose murals decorate the Town Hall.

Opposite:

Symphony concert, Christchurch Town Hall. New, and one of the world's great places for hearing music, the Christchurch Town Hall auditorium, as massed choirs and orchestra perform Mahler's 'Resurrection' Symphony.

Below:

Miles Warren, the Christchurch architect who designed the Town Hall.

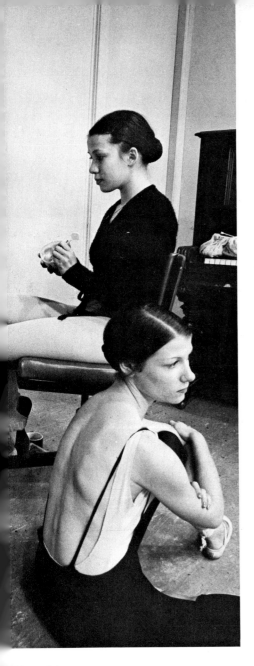

Above left:
John Hopkins rehearsing the New Zealand Broadcasting Commission Symphony Orchestra.
A former resident conductor, John Hopkins, international musician, returns for a guest performance.

Above:
'Smoko' time, New Zealand Ballet Company studios.
Icecream, coffee, sewing, knitting, thinking. Even ballet dancers must take time out for common tasks.

Opposite:
Lunchtime drama. Mercury Theatre players perform in the Auckland City Art Gallery.

They flock to theatre-restaurants like Auckland's Mercury and Wellington's Downstage, symphony concerts at the splendidly-designed new Christchurch Town Hall, arts festivals like those at Palmerston North which in recent years has produced two outstanding 'musical plays', *Earth and Sky* and *Under The Sun* by composer Jenny McLeod. Artists like Colin McCahon, Bill Sutton, Pat Hanley, Ray Thorburn, Ralph Hotere, John Drawbridge, Don Binney and Michael Smither, sell thousands of dollars worth of paintings if they hold an exhibition. The public as a whole has applauded the first system in an English-speaking country to pay writers a royalty on their books held in library stacks which, while it may not encourage writers sufficiently to become full-time professionals, will at least give them something like a reasonable reward for their writing.

New publishing presses are springing up across the land in a re-creation of the days when New Zealand had a newspaper serving every isolated farming community with a vigorous and forthright journalism. The New Zealand Press is practically the only cultural medium not yet benefiting from the country's new mood. It remains deficient in political idealism, comment relevant to its contemporary society, or leadership in human affairs. Consequently young journalists are joining a 'counter-culture' movement which almost daily proves itself by the concerns of its Press and the quality achieved in its writing and art work. Many new book publishers are joining in its wake and winning good rewards from New Zealand readers who, though they have been amazingly good buyers of 'New Zealand books', now demand something more than the tired formulae of old-established publishers. They remain avid buyers of technical and 'how-to-do-it' books which now barrage them with new ideas on how the good life, in all its dimensions, can be lived.

One phenomenon of the new culture has been Sam Hunt, the beer-mottled, private-school educated balladeer of the counter-culture, an eminently successful poet whose standards of writing are both professional and in the longest traditions of the poet's art. He sings or recites his poems at rock concerts—as well as having them published by the country's most enterprising new publisher, Alister Taylor. Tens of thousands of New Zealand's young people have had, through them, their first meaningful contact with poetry.

This is not to say that the New Zealand education system is deficient when it comes to helping children and young men and women to understand the deeper and more creative things in life. Both Primary and Secondary School teachers have, over the past twenty years, carried out a thorough blitz on the country's educational ideals and with great success. Their professional standards and idealism have prepared a people for their contemporary flowering.

Opposite:
George Henare, actor, Mercury Theatre, Auckland.

A New Zealand theatre by New Zealanders. George Henare plays the Chieftain in '*Mr King Hongi*' at Auckland's Mercury Theatre—or any role English drama demands.

Left:
Contemporary jade. Greenstone paperknife by Cliff Dalziel.

In ancient tradition, ornamental jade stone is used for a purpose, now cutting paper. A hint of the *tiki* man symbol.

Below:
Austen Deans, artist of the Southern high country.

In a long tradition of landscape art painters like Austen Deans have created a special expression of New Zealand creativity.

Bottom:
Joan Morrell, sculptor, Wanganui.

The head in bronze of arch poet, lay priest and inspirer of dispossessed urban youth, James K. Baxter (1926-1972) with friend.

SPORTS

When Billy Wallace ran up 230 points for the All Blacks during the 1905 tour of Great Britain and France he set a standard of New Zealand sporting achievement which would reach out far beyond the boundaries of the Rugby pitch. No touring All Black has ever scored more points. Wallace lived on into his eighties to become a symbol of the spirit of New Zealand Rugby. Dozens of men followed him, their names becoming part of their country's mythology. The New Zealanders' fascinated involvement with the Rugby Union game remains mysterious. Why was the game chosen by a race of people? Why did they become so peculiarly adept at it? Whatever the answers the game has played a large part in forging a sense of nationhood and it has satisfied a national sense of competitiveness. Probably it was ideally suited, as a form of community play, to a pioneering situation in which small settlements developed ways of getting to know each other.

Competitiveness in social situations has long characterized the New Zealand society. Learning to sing in New Zealand used to be regarded as a sport. Throughout the land the annual 'competitions' —the local Competitions Society's test of skill in singing, piano playing, and dancing in various ethnic styles—have traditionally been an annual event for which tens of thousands of children worked hard every year. Their successes have been great, resulting in the first successful effort to stage ballet and operas by New Zealand companies with a fair degree of public support and involvement. Talented musicians like the Auckland Catholic convent teacher, Sister Mary Leo, have produced singers of world class such as Kiri Te Kanawa, the Maori soprano who now has international recognition as a Covent Garden Opera soloist.

Such successes are part of a new mood of professionalism and the achievement of excellence in New Zealand. Just as Competitions Societies eventually produced great musical performers, the small town Bible Class harrier groups began to produce great runners and the careful tending of pastures began to produce great race horses which have made the bloodstock lines recorded in the New Zealand Stud Book among the world's best.

New Zealand's emergence as a great sporting nation on both amateur and professional levels owes itself primarily to the welfare state.

Father and son cricket match, Medbury School.
Play up, play up and play the game. Sons of gentlemen play cricket at a preparatory school sports day.

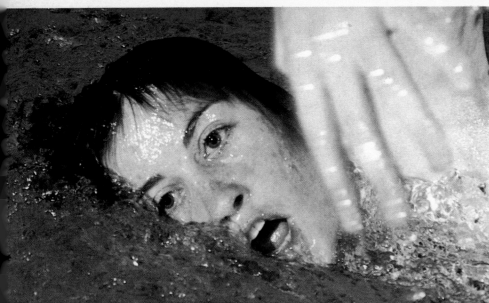

Above:

A trout leaps from the water at Rotorua, a delight to the heart of any fisherman.

Above left:

Dick Tayler from Timaru wins the 10 000 metres Gold Medal Christchurch Commonwealth Games.

The black singlet and the silver fern have clothed some of the greatest runners in human history.

Left:

Janie Parkhouse, 440 metres Butterfly Gold Medallist, Christchurch Commonwealth Games. Putting New Zealand women's swimming in the world's record books.

Opposite above:

Watching a test match, British Lions V. All Blacks.

All happy, all intent, all involved in their country's national sport, Rugby Union football.

Opposite:

Marlborough brutalise Canterbury in a rugby match that wins for them the Ranfurly Shield

The annual report to Government of the Department of Health in 1928 revealed among other things that 22 per cent of coal mining families could be classed as poor. Of timber workers' families 30 per cent were poor and 26 per cent lived in damp houses. Even in thriving farm communities mothers had to work too hard to care properly for their families—18 per cent of the houses were dirty, six per cent damp, nearly a third of children had insufficient sleep, late meal hours, hard work on the farm after school, and parents who were in fact poor. In outback farming areas 50 per cent of families were poor, 83 per cent had no bathroom, 10 per cent of children were below national average height and weight. Sharemilkers' families on dairy farms reflected the same conditions with a third of houses dirty and half of those damp, children had hard work and insufficient sleep, half the families were poor. A staggering 21 per cent of children in thriving farming communities were classed as 'backward' at school and in sharemilkers' children the figure was 26 per cent. A child was classed as 'backward' if he were *three years older* than the average age of his class. The Health Department report of 1921 had stated that 79 per cent of New Zealand children examined—scores of thousands—had a physical or mental defect of some kind and only 2-3 per cent had perfect sets of teeth.

All these things were to change and New Zealand would become a nation of champions under the welfare state with its health services, milk for school children, family benefit payments and effective pensions. Between 1935 and 1950 the infant mortality rate dropped from 32.26 per thousand for Europeans to 22.75, and for Maoris (though the figure was still shockingly high), from 109.2 to 69.74. Between 1934 and 1954 the average height of a 12-year-old girl rose by 76.2 mm and for a boy by 50.8 mm. Fifteen-year-old boys' average heights rose by 101.6 mm and girls by 25.4 mm. Girls at 12 were 2.7 kg heavier and boys 6.35 kg heavier. At 15 the twenty-year difference was quite staggering—girls averaged 7.7 kg heavier and boys 11.7 kg or nearly 2 stone! That meant an even 9.75 kg of human resources gained in every individual for the society which had demanded its freedom. It meant bigger brains, bigger hearts, more muscle, more skill, and an infinitely greater energy and sense of enjoyment of life. It meant people who were more creative and could apply their creativity to a broader range of human problems, not the least of which involved how to run faster, swim faster, sail a boat faster, train horses better, shoot sharper, kick a football further, or drive a racing car faster and in finer tuning.

There was one further ingredient necessary to release the new energies of the New Zealanders and that was a change in their education system. In 1911 the Government had given notice that 'every male inhabitant of New Zealand who on the first day of March had attained the age of fourteen years or upwards, but has not attained the age of 21 years, and who is a British subject and has resided in New Zealand for at least six months, IS HEREBY REQUIRED before 7 p.m. on the second day of June 1911 to fill in a prescribed form of registration in respect of military training under the Defence Acts and to post or deliver the same to the area

Above:
Football practice, Waitara.
Playing Rugby in bare feet, a solid training for any aspiring All Black. The great Don 'The Boot' Clarke could kick goals barefoot from beyond the halfway mark.

Opposite:
On the wicket, Christchurch Boys' High School.
Virginia creeper, 100-year-old stone, the centuries-old ballet of English cricket.

Above:

Gallop race, Riccarton. Bloodstock from family of world-beating racers.

Left:

Jockeys move out to the saddle paddoc Riccarton.

Watching and waiting, the ageless atmosphe of the turf.

Above right and left:

Spring fashion at the Races

Hat sales blossom in the weeks before classic horse race.

Opposite:

Yearling Parade, Trentham bloodstock sale

The Emperor Franz Joseph, in his Austria Imperial palaces, could not have wished f horses more classic and exquisite.

Above:
Chris Jones, farmer and polo player of South Auckland.

Opposite:
Polo players at Hamilton.
As New Zealand horses breed to ever-higher standards, thoroughbred stamina is put to the test in an increasingly popular sport.

Below:
Teeing off on a frosty morning.
Large numbers of New Zealand women of all ages are able to play sport on uncrowded weekdays.

Sergeant-Major. . . . '

While the bodies of boys born and grown on 'earth's greenest land' remained stunted they were prepared for inundation by gunfire in the Flanders Fields of the Great War.

It is remarkable how good health and physical satisfaction will change a nation's idealism and New Zealanders involved themselves in World War II with a growing sense of recognition that they really no longer belonged in Europe and had found a place of their own. It may not be going too far to suggest that the German-oppressed Mediterranean people who so befriended New Zealand soldiers in the early years of the Second War did so because they could reach a common sense of rebellion against their peasant backgrounds. And it can hardly be disputed that the New Zealanders who formed the Third Division in that war to go and fight the Japanese in the islands north-east of Australia did so with the enormous will of a people who understood that their real concerns lay in the South Pacific. The Third Division was denied early propagandist support from the New Zealand Government. It first had to recognize that there was a strong swing of public opinion away from committing troops, as Britain wished, to the terminal campaigns of Italy, Austria, and France towards defending the country where it really mattered—in the Polynesian lands.

The effects of war on the New Zealander, quite apart from his struggles to provide himself with an adequate diet and public health system, have been monstrous and pervading. Under the militarist policies of his Government at the outbreak of the Great War he went to 'the mud of Flanders' and died heroically. In the terrible destruction of Field Marshall Haig's assault on Passchendaele, 640 New Zealanders died and 2 100 were wounded, all in a morning, and in an assault which must rank amongst the most inept ever launched by a British army.

In the eight months of Winston Churchill's fateful Great War dream, the assault against the Turks at Gallipoli which guarded the Black Sea entrance straits, 2 721 New Zealand soldiers died. The wounded numbered 4 752 and some of these died later. In all, more than 18 500 New Zealanders died through the Great War and more than 50 000 were maimed or wounded. These men came from a total population of 1 100 000 people at the start of the war. The Second World War killed 11 600 New Zealanders, wounded 16 000 and saw more than 7 000 spend the best years of their life in enemy prisons. They came from a population of 1 600 000 in 1939. In such ways the emergence of a strong and inventive race in these islands was inhibited.

Shifts in emphasis and disengagement from northern hemisphere militarism have had profound effects on New Zealand. They resulted in an eventual 'de-militarization' of the New Zealand school system, first at primary level and then in the high schools which progressively asked to abandon the idea of compulsory military conscription through the half-day-a-week uniformed military instruction in their curricula. The result was that the idea of regimentation in general died in the New Zealand education system. Children continued to strive in the competitive idealism

Opposite:

'Outward Bound' style sail training, Spirit of Adventure, *Auckland*.

A $200 000 gift to a nation's youth, Mr Lew Fisher's schooner *Spirit of Adventure*, used year-round to teach young men and women about their human potential.

Above top:

Finn class yachts racing, Lyttelton Harbour.

When Brett de Thier took his Flying Dutchman yacht to the Tokyo Olympic Games to win a Gold Medal, he took also a training in Finns, the toughest single-handed class.

Above:

Big game fishing off Cape Brett.

Just another 200-pounder. A striped marlin is brought to the gaff.

of their parents but they were at last given freedom to develop and express individual talent and experience their freedoms day-by-day.

'Kicking the leather around' has always been the sport of the New Zealanders, meaning that they play Rugby Football. They have 'kicked the leather around' for a hundred years or more, traditionally with their national team the All Blacks beating all-comers including England the inventors of the game, and making arch enemies of the Springboks, the South African national side, and the Welshmen who have continually fielded tough men from the mines big enough, fast enough, and brutal enough to regard the lads from 'the Land of the Fern and the Tiki' with real ire.

Rugby players were born in bush towns. They worked all day at felling trees, shearing sheep or skinning them in freezing works, building houses or roads, milking cows, or studying hard at their professional examinations at the Universities. From the 1880s on, when small towns grew up round dairy factories or farm supply centres in New Zealand, the 'football' paddock was levelled and grassed and the towering goal posts erected before the community hall or even the church. Rugby players became New Zealand's first heroes. At Sunday School social 'spin the knife' games girls would be asked to award prizes to 'the boy most likely to be an All Black'. Newspapers would pick up stories about the new lightning fast 'winger' from Waikikamoukau and create a cult-hero out of him. A full-back who could kick goals from twenty metres further out than any man in the country would reach national prominence overnight, be selected in the North Island versus South Island annual match, be picked for the Possibles in the Probables versus Possibles All Black trials for a new tour of the British Isles, South Africa, or Australia and win through to All Black selection and a revered place in the test side. His selection match in any provincial capital city would be watched by up to 50 000 New Zealanders all baying their praise for his slightest run. After the Saturday afternoon match that same citizen would go home thinking about how his tiny son would fare on next Saturday morning's 'under tens' or 'under twelves' primary school competition match, without a thought for the icy winds which might sweep over the paddock as he stood his morning vigil on the sideline and dreamed of having bred a champion.

It has been the same with horse racing, New Zealand again producing great athletes through a combination of the bloodstock they have imported and bred from, the fine pasture lands they have developed on their stud farms, and the precision of their training methods. In the annual Yearling Sales at Trentham, near Wellington, more than $4 000 000 worth of potentially unbeatable racing talent changes hands, the annual return from a multi-million dollar investment by many of the world's wealthiest racehorse owners. The industry is paid for by ordinary New Zealanders who invest money in racing clubs and in the Totalisator Agency Board, a Government institution which controls gambling on horses and helps to supply the necessary income for a thriving bloodstock industry.

Racing horses has become 'fully professional' in New Zealand. In the last two decades New Zealanders have become professional

Rodeo competitors Lake Rerewhakaaitu.
With boots, belt buckles and bruises, straight out of an American Western tradition, a hobbling buckjumper from Antipodean 'Marlboro Country'.

Above:

Skifield, Coronet Peak, Queenstown, an international resort.

Left:

Taking a break at Coronet Peak skifield.

Opposite:

Climber abseiling rock face, Southern Alps. New Zealand climbers move on from Everest. Using new techniques made distinctly their own, they have scaled the world's most difficult mountains. A new kind of pioneering.

Below:

Harold Jacobs, mountaineer with a new technology.

experts in many other areas of sport. It is only lately that the full effects of the new New Zealander, girded in good health, a fine frame and a keen brain, have been discernible in his country's sport. As in the arts, New Zealand, quite suddenly, has spectators in sport instead of envious amateurs on the side lines. The development has coincided with similar developments in other English speaking countries whose standards of living have risen quickly. Suddenly rugby football is both a spectator sport and an enthusiast's sport instead of one in which every man child was required to be proficient. The full effects of the revolution in education, and public health, and the decay of regimentive systems in local communities co-incident with the rejection of militarism, are apparent. The Rugby Football Union administration has had to start training players who can please crowds as well as win games against international sides. If All Black sides in the old style are getting soundly beaten New Zealanders know that their own rising standards will see them through.

After all, their talents are expressed every day through the achievements of individuals like Bruce McLaren and Graeme McRae in motor racing, Peter Snell and Murray Halberg and their trainer Arthur Lydiard in athletics, Chris Bouzaid and his crews and yacht skipper colleagues who have been international One Ton Cup winners in yachting and put themselves at the forefront of the world's ocean yacht racing, and many others.

It will take a long time before the fame of the 'colonial' footballers and the racehorses like Phar Lap, ironically enshrined in a Melbourne museum, dies in the memories of New Zealanders, but they are assured that a new era of grand accomplishment has barely begun.

Opposite:
Surf Life-saving Carnival, Sumner.
Endless beaches, wild surf, restless youth, dedication, combine in a disciplined sport that builds bodies, as well as saving lives.

Below:
Water skiing on man-made Lake Karapiro, Waikato River hydro-electric scheme.
A new sport for an island people historically slow to have taken to water for amusement.

ACKNOWLEDGMENTS

The carefully planned and detailed photographic coverage for this book required a substantial financial commitment by our company, Robin Smith Photography Ltd. As so often happens, the expenditure exceeded our allotted budget and perhaps would have been beyond our resources had it not been for the generous backing given us by Air New Zealand, National Airways Corporation, Comalco, and Mount Cook Airlines.

In addition we wish to acknowledge the valuable assistance given us by Alpine Helicopters, Robin Campbell, Lakeland Aviation and N.Z. Tourist Hotels Corporation. We are also indebted to innumerable companies, organizations and persons from all walks of life whose courteous help, advice and indulgence in our objective to portray the character of the nation has contributed so much. We should like to thank our colleague, Lloyd Park, for the photograph of the Wine and Food Society on page 119, and Eric Bell for the photograph of the Stone cottage in a hoar frost on page 28.

                              Robin Smith   Warren Jacobs

PHOTOGRAPHIC DATA

Consideration of photographic problems likely to confront us on this assignment led us to realize that we would need to choose cameras of the utmost versatility. Colour transparencies large and sharp enough for the faultless production of double-page spreads were essential. But these requirements had to be compatible with cameras light in weight and fast in operation with a wide variety of lenses.

The equipment which most closely met these requirements was the Rolleiflex SL66 camera with its range of superb Zeiss lenses. We used four of these cameras with two 40mm, two 50mm, three 80mm, two 150mm, two 250mm and one 500mm lens. This standard equipment we backed with two Nikkormat FTN cameras using lenses from 20mm to 400mm and one magnificent Leicaflex camera and Leitz lenses. These 35mm cameras were used for the purposes for which they are so ideally suited . . . split-second action on the sports field and in low lighting situations which occur frequently with dance and drama productions. Here the small cameras came into their own with ease of operation in difficult conditions, the quietness of their shutters, their unobtrusiveness and capability of use with slow shutter speeds while hand-held.

Several pictures were taken with large format cameras. Linhof 5x4s with Schneider, Zeiss and Rodenstock lenses from 65mm to 360mm were chosen here. In addition, we used Rollei E 36 RE, Mannesmann and Mecablitz electronic flash units where the lighting situation demanded them.

There were occasions when we wished for a wider angle lens or an even stronger telephoto but the equipment which we had purchased met virtually every requirement and, in retrospect, our choice would remain unaltered. To the manufacturers and suppliers of this equipment we offer our compliments and thanks.

                                                    R. S.
                                                    W. J.